JOHN T. HARDY

American University of Beirut

SCIENCE
TECHNOLOGY
and the
Environment

 *SAUNDERS GOLDEN SERIES
IN ENVIRONMENTAL SCIENCE*

1975   **W. B. SAUNDERS COMPANY**
Philadelphia, London, Toronto

W. B. Saunders Company: West Washington Square
Philadelphia, PA 19105

12 Dyott Street
London, WC1A 1DB

833 Oxford Street
Toronto, Ontario M8Z 5T9, Canada

Library of Congress Cataloging in Publication Data

Hardy, John T

Science, technology, and the environment.

(Saunders golden series)

Includes index.

1. Science. 2. Technology. 3. Science — Social aspects.
   4. Technology — Social aspects. I. Title.

Q172.H37   301.24'3   74–11688

ISBN 0–7216–4513–5

Photograph courtesy of A. Jack Swern.

Trademark courtesy of Hiram Walker and Sons, Inc.

Science, Technology and the Environment   ISBN 0-7216-4513-5

Last digit is the print number: 9   8   7   6   5   4   3   2   1

DEDICATION

To Sheila my comrade in time and Kevin
of the coming generation.

PREFACE

The century would seek to dominate nature as it had never been dominated, would attack the idea of war, poverty and natural catastrophe as never before. The century would create death, devastation and pollution as never before. Yet the century was now attached to the idea that man must take his conception of life out to the stars. . . . A century devoted to the rationality of technique was also a century so irrational as to open in every mind the real possibility of global destruction. It was the first century in history which presented to sane and sober minds the fair chance that the century might not reach the end of its span. It was a world half convinced of the future death of our species yet half aroused by the apocalyptic notion that an exceptional future still lay before us. So it was a century which moved with the most magnificent display of power into directions it could not comprehend. The itch was to accelerate — the metaphysical direction unknown.

Norman Mailer
Of a Fire on the Moon

Many people now question the role of science in promoting man's welfare. Research funds, relatively abundant for the past 15 to 20 years, are once again becoming scarce. Science education has not responded to current changes in the motivations and drives of contemporary students, who increasingly turn to other fields of endeavor. Professors of science, their own vision often restricted within a narrow field of research and teaching, remain ignorant of the broader implications of their work and frequently find difficulty in conveying the "broad picture" to nonscientists. Yet the general public must be well informed about science and technology, so that they may grasp, participate in, and control the direction of the scientific and technological society in which they now live.

In conveying scientific information to nonscientists I make several assumptions. First, curiosity is one of man's basic characteristics; thus, nature is interesting even to nonscientists. The historical development of scientific ideas, the present state of flux of these ideas, the role of science in natural philosophy and in influencing man's future, all make the study of science of utmost interest. Second, scientific knowledge in its basic precepts is comprehensible to those not trained in technical or mathematical knowledge. Both the general esthetic appreciation of scientific knowledge as well as the inherent dangers of misapplied technology can be grasped by the nonscientist without need of technical details.

This book, directed toward nonscientists, is based upon a freshman and sophomore university course entitled *Contemporary Topics in Natural Science* — actually a course in the appreciation and understanding of science, analogous perhaps to contemporary courses in art or music appreciation. No attempt is made to "cover material," because the range of scientific knowledge cannot be contained within the covers of a book; rather, the general background of some of the most thought-provoking contemporary problems of science and technology are presented. It was possible only to touch on the general nature of some of these problems, but references at the end of each chapter are largely to non-technical works that will provide stimulating further reading for interested students.

What will the twenty-first century be like? We live in a scientific-technical world. Science fiction often provides lucid insights into the future direction of this world. Many of the imaginative creations of science-fiction writers seem to have a way of rapidly becoming science fact. In 1868 Jules Verne described man's journey to the moon in a rocket ship — an event that today hardly makes headlines. Deadly bombs, germs, and chemicals, all utilized in the wars of societies depicted in science fiction, are today possibilities to be reckoned with. The deadly but fictional ray gun of Buck Rogers and other early science-fiction characters exists today (high energy laser beams), and is being developed as a weapon of war. An early and recurrent science-fiction theme — the machine, electronic robot, or supercomputer that controls rather than serves man — is certainly a modern possibility, if not in some cases already a reality. Forty years ago Aldous Huxley's classic *Brave New World* predicted many of the questionable uses to which genetics, embryology, and psychology can now be put. An increasing number of science-fiction authors now turn their attention to the problems posed by modern psychobiology. For example, Michael Crichton recently provided a poignant description of the dangers and implications of modern surgical-electronic mind control in *The Terminal Man*. Thought by many at the time (1798) as overimaginative fiction, Thomas Malthus wrote of the impending collision between the world's growing population and the limited food supply; and later writers such as John Brunner (*Stand on Zanzibar*) correctly portrayed future cities as crowded, sprawling, multilayered megalopolises where all available resources had to be recycled and used efficiently.

Finally, many science-fiction writers, from Aldous Huxley (*Brave New World*) to George Orwell (*1984*) and Ray Bradbury (*Fahrenheit 451*), foresaw in the application of modern scientific technology an emerging loss of individual freedom. As if to confirm their predictions, behavioral psychologist B. F. Skinner argued in his recent nonfiction book (*Beyond Freedom and Dignity*) that this individual freedom is simply an illusion, that it has never existed. Are science and technology leading us away from a simpler freer life toward a world "beyond freedom and dignity"? Or, on the other hand, will the benefits of scientific knowledge pave the way for world affluence, peace, and growing individual freedom? The following chapters review current knowledge from relevant areas of the physical, biological, environmental, and social sciences, and examine the implications of each field for man's freedom and his future in general.

I believe an understanding of science and technology is of greater importance and relevance today than ever before. Our world confronts us daily with problems that demand scientific knowledge and technical expertise coupled with humanism. Society cannot afford to make decisions based only upon outmoded knowledge that lags behind today's complex technology. We must convey an appreciation of the proper role and uses of scientific knowledge to students who, in the future, will have these powerful means at their disposal. In order to direct science and technology for the benefit of mankind, nonscientists must understand the origins, present state, and future promise or peril of science and technology. As Gerald Holton stated in 1960, "To restore science to reciprocal contact with the concerns of most men—to bring science into orbit about us instead of letting it escape from our intellectual tradition—that is the great challenge that intellectuals face today."

Many individuals aided in the creation of this book. In particular, I thank Dr. Peter Frank (University of Oregon) for his careful review of and suggestions on the entire manuscript. Invaluable comments and criticisms on particular chapters were provided by Drs. Lawrence Edwards (American University of Beirut), Garrett Hardin (University of California, Santa Barbara), E. W. Pfeiffer (University of Montana, Missoula), J. B. Neilands (University of California, Berkeley), Fred Leavitt (California State, Hayward), Amos Turk (The City College of the City University of New York), Janet Turk Wittes (Columbia School of Public Health), Phillip Basson (American University of Beirut), William

Hayes (University of Georgia), Roger Del Moral (University of Washington), Russell Christman (University of North Carolina, Chapel Hill), and Robert Fischer (California State College, Dominguez Hills). I also wish to express appreciation to the staff of W. B. Saunders Company for their competent editing and production and to Ms. Huda Cook for typing the manuscript.

CONTENTS

1

THE NATURE
OF SCIENCE,
TECHNOLOGY,
AND SCIENTISTS

...scientists are becoming aware that they can no longer claim that the pursuit of knowledge is divorced from its use. Yet, to face the responsibility for the application of their work would impair their freedom. The loss of freedom is part of the dilemma of science.

Anthony Michaelis and Hugh Harvey

Science and technology have played major roles in the development of modern industrial countries. In 1974, the United States budgeted approximately $17.4 billion for research and development: $9.4 billion for military, $2.5 billion for space and $5.5 billion for civilian. This represented about 6.25 per cent of the total national budget. Support of science and technology is strong, and technology has played a major role in shaping modern society. Science coupled with technological innovation has certainly contributed in many ways to the material well-being of the peoples of the world. Yet there are those who now question the importance of pursuing further scientific and technological growth. They see in science not the savior but the destroyer of worlds. Thus it becomes imperative that we examine the present role of science and technology in our society and elucidate the promise or peril that it may hold for the future of mankind.

WHAT IS SCIENCE?

Science is an attempt to make the chaotic diversity of our sense-experience correspond to a logically uniform system of thought.

Albert Einstein
Out of My Later Years

1

The word *science* is derived from the Latin *scientia* meaning "knowledge." Science, of course, is more than just knowledge, and the German word *Wissenschaft*, meaning "organized knowledge," comes closer to the meaning of science. But surely part of our definition should also include the method used by science. This method is observation. Scientific observations are usually systematic; that is, they are organized in a meaningful way. R. B. Fischer has defined science as *an organized body of knowledge based upon methods of observation.*

Natural science, as used in this text, refers to basic knowledge in the physical, biological, environmental, and behavioral sciences. Science is knowledge gained through experience; this experience, as the British empiricist philosophers pointed out, is obtained through the senses of touch, smell, taste, hearing, and vision. Science is based upon observation and is limited to those things which can be observed by the human senses. The human senses can be extended by the use of instruments such as telescopes and microscopes, but ultimately we must rely on our senses for knowledge of the natural external world. Science does not deny the existence of supernatural or mystical phenomena (those things which cannot be observed by the known senses) but simply states that such phenomena are not amenable to scientific or observational inquiry; that is, they are outside the scope of science.

Scientific Methods

Scientific methods include the formation of hypotheses, observations, generalizations, explanations, and predictions. Science involves the collection of data about facts. A fact is a natural occurrence of an event. A *datum* is the recording of that event and is often represented by some symbolic form such as colors, sounds, pointer readings, or weights. Data thus describe facts in the language of physical terms.

Immediate experience on what may be called the *perceptual (sense) plane*, if uninterpreted and uncorrelated, does not extend the realm of science. The scientist, like the layman, associates different perceptual (sense) experiences to formulate ideas or concepts called *constructs*. To achieve this association, some means of measurement is often used (Fig. 1–1*A*). For example, by quantifying experience in numerical values, the experience can be described in terms of logical thought. The tran-

Figure 1–1   The size range of objects in the universe is tremendously large. To express such measurements, scientists use powers of 10, often called *scientific notation. (For an explanation of such numbers see Appendix.)* (From Parsegian, V. L., et al.: Introduction to Natural Science. Part I: The Physical Sciences. New York, Academic Press, 1968.)

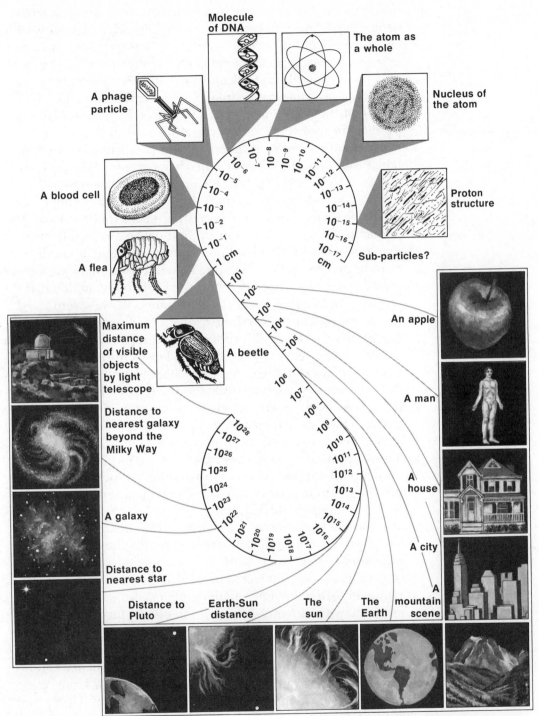

Figure 1–1 *See opposite page for legend.*

sition from sense experience to concept is made in specific steps by the method of *induction*, by which one progresses from particular experiences to generalizations. The scientist begins with observations and proceeds by rules of correspondence to a concept (construct) that can be related logically to other constructs. He may repeatedly go back and forth between the perceptual (observational) and conceptual (generalizational) planes; but eventually he arrives at a set of constructs that enable him to predict future events within certain statistical limits.

Validation and Verification. If his predictions seem to be borne out, then the scientist's theory is considered feasible, but it is always subject to further testing and confirmation or revision. Scientific constructs at the conceptual level must be verifiable on the perceptual level by critical experiments and deduction. A scientific fact or event must be observable by more than one scientist. Others under the same or similar conditions must be able to deduce the same fact; it must have *intersubjective testability*. It must meet the criterion of reliability, so that different people agree upon the description of the event.

Scientific constructs should have a maximum measure of extensibility and logical fertility; they should explain more than a single type of event or occurrence. For example, Newton's Universal Law of Gravitation (see Chap. 2) explained not only the fall of an apple from a tree, but the fall of all objects near the surface of the earth as well as the motions of the moon, earth, and other planets about the sun. Acceptable constructs should also provide connections to other sets of constructs. Thus, science obtains a set of observed facts, relates them to each other in logical patterns, formulates constructs that approximate these relationships, and extends predictions to be tested by further observations.

Basic scientists, then, are engaged in the task of articulating, deriving, and generating principles that have general explanatory power. Their work may be quite narrow and consist of supplying a specific detail of an already existing theory. However, often the most challenging, exciting, but rarest of scientific activities involves the creation of new principles and theories which lead to large-scale modifications of previously held concepts.

Most basic scientists are associated with universities and receive financial support from teaching positions or from research grants, often from the federal government. Unlike the applied scientist (see later in this chapter), the basic scientist generally does not work on a limited time scale and in some cases does not even know in what direction an interesting problem may lead, what the solution will be, or whether he will ever arrive at a solution. The seventeenth-century German astronomer Johannes Kepler spent more than 30 years formulating his Three Laws of Planetary Motion. Charles Darwin's publication of *On the Origin of Species* came more than 20 years after his gathering of observational data during the voyage of the HMS *Beagle*.

Scientific progress depends upon both inductive and deductive reasoning, working back and forth and often involving many small steps. However, progress may also be made in larger steps by very creative imaginations and by what has been called intuition—an uncanny ability for contemplating a set of facts or observations and from there to take an "inductive leap" over intervening steps of logic or experiments to a valid set of concepts or ideas.

The Uncertainty of Science

Science seeks "truth." The word *truth* is difficult to define. Its meaning has been debated by philosophers for centuries, but as generally used in science, truth has a statistical meaning, and scientists recognize that they can never determine exactly or absolutely the truth. However, by carrying out experiments with instruments of progressively greater accuracy or by repeating experiments, they arrive, so to speak, *closer to* the truth. There are limits to scientific understanding, and scientific truth is probabilistic.

There are two ways by which one can know with certainty whether a statement is true. One can define it as true, or it can be derived logically from statements that are defined as true. For example, $2 + 2 = 4$ is a truth derived from the definitions of whole numbers, but true statements about classes of events in the natural world cannot be derived from definitions alone. Any law or explanation in natural science ultimately rests on *induction* from past observations, and future predictions cannot be absolutely certain. In other words, even if we observe an event on several occasions under similar conditions, with high regularity, we cannot predict with absolute certainty that the event will occur in the same way again. For example, we expect that the sun will rise every day; we have been told that it will; we understand the scientific explanations of this occurrence; and we know that it has risen every day of our lives. However, these reasons are not sufficient to enable us to predict *for certain* that the sun will rise tomorrow.

Not only can we not be certain that the same situation will lead to the same results, but in actuality, the same situation never occurs more than once. In the natural world, many events occur simultaneously, and the scientist, in order to sort out the relationships among them, sets up conditions to minimize the number of variables. He attempts to control the conditions under which experiments are carried out by observing the sequence of events under artificial but well-specified conditions. Nevertheless, in repetitions of the same experiment, conditions are never the same. For example, if we wish to measure the boiling point of water, we put some water in a beaker, insert a thermometer, and apply heat until it boils. If we do the same thing the following day we use different water; the beaker is different or at least is now a day older and has already been used to boil water; the thermometer is also older and previously used; furthermore, the positions of the sun and moon with respect to the earth are different on the second day. We *assume* that these conditions will not make a difference, but we do not *know* this with absolute certainty. Each situation is unique.

How long is a year? Conceptually a year is the time required for the earth to make one revolution around the sun. In the past, this was *estimated* to be approximately 365 days. Later, it was estimated to be closer to 365¼ days. In order to prevent the calendar year from falling behind the solar year, one extra day was added every four years. This *estimation* of the length of the year came closer to what we might consider truth. However, with more precise instruments, namely, the atomic clock (see Chap. 2), it has become possible to estimate the actual time of one revolution of the earth around the sun with extreme accuracy and now we add not only one day every four years but one day and one second (the leap second).

The probabilistic nature of science can be seen in the following example from McCain and Segal:

In all of science, probably the one measure that currently stands as the most absolute is the constancy of the speed of light; however, the best estimate of this constant, though it is very good, changed from 299,776 kilometers per second in 1941 to 299,792.8 kilometers per second in 1955. The numbers may look very similar, but they differ by 37,584 miles per hour. Furthermore, Bearden and Thomsen, who measured the speed in 1955, were only 50% sure that they were within 896 miles per hour of the "true" velocity.

Another constant in an "exact" science is Avogadro's number. This is the number of molecules in one mole (gram-molecular weight) of a substance. The number is given as 6.02486×10^{23} molecules, but there is a 0.0027% error in this estimate. That is a very small error, but it comes to about 16,000,000,000,000,000,000 molecules. That is, Avogadro's number is accurate to about 16 quintillion molecules.

What Are Scientific Laws?

Scientific laws are statements that describe the properties of similar events or sequences of events. The so-called laws of nature are derived and constructed by man to describe how nature operates. They have been established after observation of consistent repetitions of interactions between events. It is the degree of reproducibility of a phenomenon that leads to its acceptance as a law. For example, the fact that any object starting from free fall in a vacuum will consistently demonstrate the equation $S = \frac{1}{2}gt^2$ * leads one to call it a law. Nevertheless, a series of measurements taken at St. Louis, Missouri produced a different set of results. We can only conclude with some degree of assurance (confidence) that 95 per cent of the measures will fall within 0.010 cm of the average (Fig. 1–2). The so-called laws of science are, strictly speaking, something less than absolute. Realizing that all such cases are approximations, however, does not diminish their predictive usefulness in actual situations.

The scientist accepts a certain range of variation. He may attempt to limit this range as much as possible, but he can never eliminate it completely. As noted earlier, scientific truths are never complete but always subject to revision based upon new evidence. Such truths are tentative. Science does not seek any ultimate explanation of nature in one all-encompassing theory, but concerns itself with limited and potentially solvable problems. It is this concern with limited and solvable problems that has led to the spectacular successes of the scientific method. There are no "exact sciences"; science is probabilistic. Mathematics is often referred to as an exact science but, although vital to science, it is really a logical tool used by scientists. Science must be related to observation; pure mathematics is based upon the logical consequences of a set of postulates. Thus, observationally based sciences are often called *natural sciences*, and mathematics and logic, *formal sciences*.

How Do Applied Science and Technology Differ from Basic Science?

The practical results of science are the most visible to the layperson and it is the public's desire for practical gain through the by-products of scientific work that often leads to the financial support of research. Such visible technological achieve-

---

*Where S = distance traveled by falling body in first two seconds;
g = the gravitational "constant," which varies with location and altitude;
t = time in seconds.

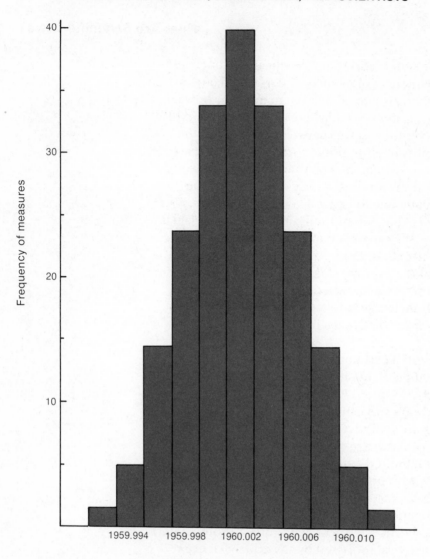

Figure 1–2 Estimated distribution of 200 measured distances traveled by a falling body in the first 2 seconds starting from rest. (From The Game of Science, Second Edition, by G. McCain and E. M. Segal. Copyright © 1969, 1973 by Wadsworth Publishing Company, Inc. Reprinted by permission of the publisher, Brooks/Cole Publishing Company, Monterey, California.)

ments as the atom bomb, the polio vaccine, and rockets and space travel have led the general public to think of science in terms of applied science — a collection of facts leading to practical ends. But such accomplishments are really the application of already existing scientific knowledge.

Although technology is often equated with applied science, technology itself has developed independently of science throughout most of recorded history. Technology does not usually include an organized and systematic series of observations but simply develops by trial and error. The invention of the waterwheel or the windmill thousands of years ago are examples of technological innovations that preceded any organized scientific knowledge of the principles involved or their means of operation. Technology is as old as mankind, and primitive man the toolmaker was the original technologist. After the scientific revolution and the emergence of an organized body of knowledge, it became apparent that much of this knowledge could be applied to increase man's material well-being. R. B. Fischer has defined technology as the *totality of the means employed by peoples to provide material objects for human sustenance and comfort.* Late in the nineteenth century, we find the development of a science-based technology; that is, applied science. Today technology is largely dependent upon science, and a two-way interaction operates between science and technology, each reinforcing the other.

All fields of endeavor, including agriculture, medicine, arts and humanities, education, and science now utilize scientific knowledge in developing new technologies. An example of this interaction between science and technology is the development of the electron microscope. A large body of organized knowledge concerning electromagnetic radiation and its properties had to precede the actual building (in 1934) of the electron microscope, which utilized this existing basic scientific knowledge. The electron microscope was then used by scientists to study the fine structure of membranes and organelles within living cells, and further increased the organized body of scientific knowledge. Similarly, the development of the polio vaccine followed a long historical development of basic scientific knowledge. The idea that diseases are the products of specific infective agents had been known since the late nineteenth century work of Louis Pasteur. In 1892 the Russian botanist D. J. Ivanovski laid the experimental foundation for research on viruses by transmitting tobacco mosaic virus using cell-free extracts of infected plants. By 1908, methods had been developed for transmitting poliomyelitis virus to rhesus monkeys, and in 1949 Enders and his colleagues managed to isolate and maintain cultures of the virus in tissues taken from human embryonic organs. In addition, basic concepts concerning viruses and principles of immunity were partially understood, and detailed knowledge of anatomy was available. All of this knowledge and more was utilized in the development of a polio vaccine by Jonas Salk in 1953.

Applied scientists employ already existing principles and concepts to solve specific and generally limited problems. In solving such applied problems, new basic scientific knowledge is sometimes discovered, but the major effort is directed toward utility rather than discovery. The activities and characteristics of applied and basic scientists often overlap and are not always clearly distinguishable.

Are Science and Technology Good, Bad, or Ethically Neutral?

In order to answer this question, let us examine the products of science and technology. The product of scientific observation is organized knowledge. This in itself is good or at least ethically neutral. However, unlike science as we have defined it, the products of technology generally consist of material objects. The uses to which these technological products are put can be either good or bad. Basic scientific knowledge about nuclear physics can be used to develop the technology for generating electrical power for the purpose of distilling salt water into fresh water for crop irrigation (thus increasing the world's food supply); or the same basic knowledge can be used in the technology of thermonuclear bomb construction (for possible destructive purposes).

Now that we live in a technological environment, the following question arises: Is man the master or the slave to this new technology? Is technology expanding his choices and freedoms or limiting them? Up to the present, man has proved capable of directing and orienting technology to desired ends; however, with the acquisition of knowledge leading to an almost absolute power over nature, man must now carefully weigh the desirability of alternative technological choices. This new power must be used for humane ends (see Chap. 10).

What Are the Origins of Science and Technology?

Science is contemporary. We are living in a modern age of science and technology that dawned primarily within our own lifetimes. This fact is well stated by Alvin Toffler in his book, *Future Shock*:

If the last fifty thousand years of man's existence were divided into lifetimes of approximately 62 years each, there have been about 800 such lifetimes. Of these 800, fully 650 were spent in caves. . .

Only during the last 70 lifetimes has it been possible to communicate effectively from one lifetime to another — as writing made it possible to do. Only during the last 6 lifetimes did masses of men ever see a printed word. Only during the last 4 has it been possible to measure time with any precision. Only in the last two has anyone anywhere used an electric motor. And the overwhelming majority of all the material goods we use in daily life today have been developed within the present, the 800th, lifetime.

This 800th lifetime marks a sharp break with all past human experience.

Although the majority of scientific and technological developments have occurred only recently, we can identify small beginnings of science and technology that occurred thousands of years ago. The ancient civilizations of Sumaria, Egypt, and Babylonia collected a great deal of observational evidence on the movements of the heavenly bodies, and ancient Greece provided much of the basic scientific thought that later served as a starting point for a revolution in the minds of men.

The Hellenistic period of Greece, from Alexander's conquest in the third century BC until the beginning of the Roman Empire, lasted 350 years. During this period, we find many of the origins of western scientific thought. The ancient Greeks emphasized a philosophy of logic and mathematics, and began to make observations of the natural world. Many of the writings of Aristotle, for example, deal with attempted classifications of animals and plants. A great center of learning was established in Alexandria on the Nile delta in Egypt, and there Aristarchus (circa 310 to 230 BC) proposed a model of the universe with the sun at the center—a revolutionary idea not accepted by his contemporary Greeks. Eratosthenes (273 to 197 BC) used the angles of the sun at two different latitudes to compute the circumference of the earth.

Many technological advances occurred during the Hellenistic period. Hero of Alexandria built the first steam engine, and Archimedes constructed various types of pulleys and levers. However, when their observations of nature differed from their preconceived ideas based upon mythology and polytheism, the Greeks often abandoned their newfound observational knowledge in order to eliminate contradictions. Finally, during the last centuries of the Hellenistic age, superstition and mysticism began to replace logical thought.

The Roman Empire, heir to this latter period of Hellenistic thought, contributed very little to natural science. Astrology became popular and Christianity arose during this period. The early Christians, attempting to rid themselves of their pagan past, sometimes destroyed temples and works of art and science. Rome, captured by invaders from the north in 476 AD, fell into economic decline and chaos. The situation in the West was not conducive to the development of the arts and sciences. The eastern Roman Empire, on the other hand, centered in Constantinople, survived this period more or less intact, and learning shifted to the East. The works of the Greek philosophers were translated into Syriac and the center of learning finally moved further eastward into Persia.

In the seventh century AD Mohammed founded a religion in the Arabian peninsula. Islam spread rapidly across the Middle East, North Africa, and Spain. This newly flowering culture assimilated much of the knowledge from the civilizations with which it came into contact, particularly the Greeks, Chaldeans, Indians, Persians and, perhaps in the case of alchemy, Chinese, and integrated this knowledge into a new scientific tradition. In the third century AD Shāpūr I founded Jundishapur at the site of an ancient city near the present Persian (Iranian) city of Ahwaz. This became a center for ancient sciences, which were studied in Greek, Sanscrit, and later in Syriac. A school was set up at Antioch on the model of the Greek schools of Alexandria where medicine, mathematics, astronomy, and logic were taught, mostly from Greek texts translated into Syriac. Knowledge from Indian and Persian sciences was also included, and the school lasted long after the establishment of the Abbassid caliphate as an important ancient source of learning in the Islamic world.

Islamic science was closely associated with its religious heritage and attempted to explain the unity of the universe and nature. Unlike later western science, Islamic science never wholly broke away from its religious authority. *Scientia* or human knowledge was regarded as a noble pursuit only so long as it was subordinated to *sapientia*, divine wisdom. Science did not establish an independence based on its own principles nor did it attempt to encompass the infinite within some finite system. Science in the western world has concentrated its energies upon the study of quantitative relationships and the development of a science of nature, whose obvious fruits have been physical technology and its applications. Islamic science, on the other hand, sought ultimately to obtain knowledge that would contribute to the spiritual enlightenment of anyone capable of studying it. Its fruits were inward and hidden and its values more difficult to discern.

For the next few centuries the Arab world formed the center of science in the western world. The Arabs demonstrated keen interest in Greek learning and made many contributions of their own, especially in the area of mathematics. However, by the fifteenth century, the Moslem empire began to disintegrate. In Europe during this period (from the seventh to the fifteenth century), religious interpretations of the universe dominated men's minds and very few advances in knowledge took place. In the Far East, however, several developments occurred during this period. Chinese mathematics became highly developed, and Chinese technology produced porcelain, improved ship designs, block printing, and gunpowder. Astron-

omy was pursued in India based upon the knowledge of the
Greeks, and the concept of zero was introduced into mathematics by Hindu mathematicians.

Between the twelfth and fifteenth centuries Latin translations of Greek works from the Arabic began to appear in
Europe. As Parsegian says, "The philosophy of Greece had
traveled from Greece and Alexandria via Rome and Constantinople to Baghdad and Andalusia and thence back
to Rome." During the twelfth century, a revival of learning
occurred in Europe. Several universities were founded
where Greek and Arabic writings were studied and interpreted. Roger Bacon composed a book of Greek grammar so
that more scholars could go back to the original sources of
science and philosophy. Nevertheless, man's view of the world
was still that presented by Dante in the *Divine Comedy* (Fig. 1–3).
However, by the fourteenth century, we find the beginnings of
a new period, a revolution in the history of science — the Renaissance of learning that radically altered man's picture of the universe.

By the year 1500, western civilization had begun to recapture the ancient Greek writings and scholarship, and we find
during the Renaissance the beginnings of the so-called scientific revolution. Interestingly, it may have been the artists and
craftsmen of the Renaissance who actually formed the small
beginnings of a scientific community. The workshops of such
artists as Leonardo da Vinci actually resembled laboratories.
Renaissance artists attempted to represent the human body in
great detail. Thus, the study of human anatomy became one of
the earliest sciences in Europe. This intense sixteenth-century
interest in anatomy revealed some of the classical Greek medical and anatomical writings to be incorrect. In England in the
year 1628, William Harvey demonstrated the circulation of
blood in the human body, and his work provided an example of
what could be accomplished by careful observation.

During the seventeenth century the telescope and the microscope opened up new worlds to man's observational abilities.
Galileo formulated a new concept of the solar system, displacing the former geocentric theory. Isaac Newton put forth his
universal theory of gravitation to explain the motions of the
planets around the sun (see Chap. 2).

During the eighteenth century there were many significant
discoveries in the area of chemistry. In England, Joseph Priestley discovered the existence of oxygen; Antoine Lavoisier in
France discovered that water was composed of oxygen and
hydrogen; Robert Hooke in England, using the newly developed microscope, found worlds of minute organisms never
before seen by man — even a drop of pond water, he discovered, teemed with tiny life. Expeditions set out from many
countries of western Europe, and ships brought back collec-

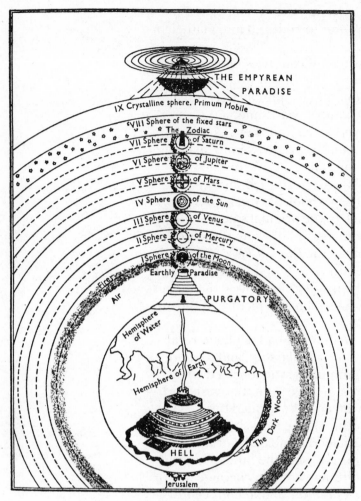

Dante's scheme of the universe

Figure 1–3 (From The Fabric of the Heavens by Stephen Toulmin and June Goodfield. Copyright © 1961 by Stephen Toulmin and June Goodfield. Reprinted by permission of Harper & Row, Publishers, Inc., and Curtis Brown, Ltd.)

tions of plants and animals from all parts of the globe. In Sweden, Carl von Linné (Linnaeus) developed a scheme of classification for plants and animals. By the beginning of the nineteenth century, scientists were aware of the tremendous diversity of living things and the stage was set for the formulation of a great biological theory.

In the nineteenth century Charles Darwin, the son of an English physician, after spending some time studying medicine, turned to the study of natural science and biology. He soon found an opportunity to sign on as a naturalist on the British exploratory ship HMS *Beagle*, for a cruise around the world. The voyage of the *Beagle* from 1831 to 1836 was, in Darwin's words, ". . .by far the most important event in my life, and has determined my whole career," for it was on this long voyage that Darwin made the extensive collections and observations on the diversity of plants and animals that later led to his theory of evolution. Darwin was not alone in his ideas on the origin of species by natural selection. Evidence on the diversity and interrelationships of organisms had been accumulating for years, and other scientists such as Alfred R. Wallace were forming the same theory. After giving a joint paper with Wallace before the Royal Society in England, Darwin continued to gather supporting evidence and in 1859 published his classic work, *On the Origin of Species by Means of Natural Selection.* Scientific knowledge, particularly in biology, continued to grow throughout the nineteenth century.

New areas of inquiry were opened up. Modern physics and relativity dawned with the twentieth century. With each gain in scientific knowledge the rate of scientific growth accelerated exponentially. New findings led to more new findings; the larger the body of organized knowledge, the faster was the rate of growth. Evidence from many sources, such as the rate of growth in the number of scientific journals and publications, and in the number of living scientists, has led to estimates that scientific knowledge is presently doubling every ten to fifteen years (Fig. 1–4). This becomes dramatically clear when we consider that 80 to 90 per cent of all the scientists that ever lived are living today. In the United States at the present time there

Figure 1–4 Total number of scientific journals and abstract journals founded, as a function of date. Note that abstracts begin when the population of journals is approximately 300. Numbers recorded here are for journals founded, rather than those surviving; for all periodicals containing any "science" rather than for "strictly scientific" journals. Tighter definitions might reduce the absolute numbers by an order of magnitude, but the general trend remains constant for all definitions. (From de Solla Price, Derek J.: Science Since Babylon. New Haven, Yale University Press, 1961.)

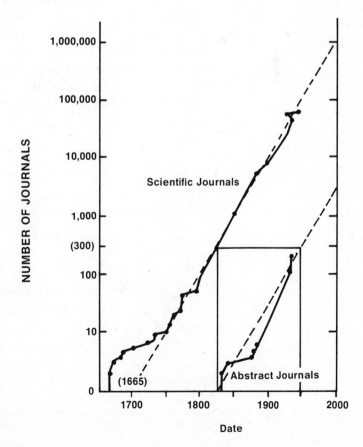

are more than one million people with scientific and technical degrees.

Can Science Continue to Grow?

Are there limits to what we can learn, or is knowledge limitless? Although diversity is great in the natural world, the universe (and thus, nature) is probably finite (see Chap. 2) and may be comprehensible to a large extent at the present time. In actuality, exponential growth cannot continue to infinity, whether it be the growth of dollars in interest or the growth of human or animal populations. In many natural cases a limit is reached, growth slackens, and a typical logistic curve results (Fig. 1–5). Derek J. de Solla Price has suggested that at some time, perhaps in the 1940s or 1950s, we passed through the midperiod in this logistic growth of science. Others have estimated from extrapolations of exponential growth that scientific knowledge will cease growing within a generation or two.

Are we now living in a "golden age" of science and affluence never to be duplicated in the future of man? In the past, as scientific knowledge grew, scientists came to understand the universality of the laws of nature: matter is composed of the same particles and elements everywhere in the universe; radiant energy moves with the same velocity and has the same characteristics everywhere. Modern biology and biochemistry have demonstrated the basic chemical similarity of all living organisms. This uniformity of nature and general applicability of natural laws sets limits on our knowledge. What remains to be learned may be staggering to the imagination, but future scien-

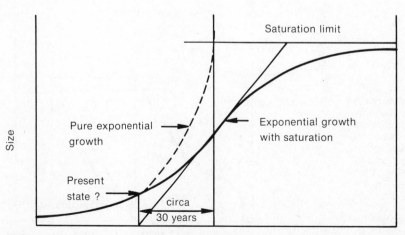

Figure 1–5 Form of the logistic curve of scientific knowledge. (From de Solla Price, Derek J.: Science Since Babylon. New Haven, Yale University Press, 1961.)

tists may spend much of their time in synthesizing and reviewing the vast body of knowledge already accumulated and in communicating this knowledge to scientists in other fields and to the public at large. Gunther Stent predicts an end to exponential growth in the arts and sciences within a few centuries. We shall then arrive at a "golden age" when the aggressive instinct will have almost vanished and a small group of workers will keep intact the technology to sustain the remainder of the population at a high standard of leisurely living.

Alvin M. Weinberg of the United States Oak Ridge National Laboratory, writing on the impact of large scale science in the United States, describes the recent emergence of "Big Science" supported by large amounts of public funds and administered by a large bureaucracy.

> When history looks at the twentieth century, she will see science and technology as its theme; she will find in the monuments of Big Science — the huge rockets, the high energy accelerators, the high flux research reactors — symbols of our time, just as surely as she finds in Notre Dame a symbol of the Middle Ages. . . . We build our monuments in the name of scientific truth, they built theirs in the name of religious truth.

Several problems arise with the emergence of Big Science. Big Science requires a large organization; it thrives on public support and on publicity. Scientists often become concerned with budgetary problems and money-spending instead of original thinking; consequently many scientists become administrators. Financially, the increase in science-related spending has been about 10 per cent per year, corresponding to a doubling time of seven years. Since the doubling time of the GNP (gross national product) in the United States is about 20 years, at this rate of growth, the United States would be spending all of its money on science and technology in about 65 years. Obviously such large growth in spending on science cannot continue. Much of the emphasis on scientific effort in recent years has been on space flight technology and high energy physics, projects which offer little immediate promise of increasing human welfare. Many scientists now question whether or not Big Science should be directed toward more relevant problems of environmental pollution, energy and resource development, and medical and birth control technology.

What Are Scientists Like?

Scientists are human beings! This may seem obvious to the reader, but let us look at two typical public images of scientists.

First, many people believe that scientists are more than human. Scientists are idealized and idolized. It is believed that they can solve any problem and that their products can provide us with all of our needs — they are superhumans. Second, and less flattering, is the view that scientists are less than human. Scientists are viewed as being cold, detached, impassive, and engaged in unconcerned observation of phenomena which have no emotional meaning for them (Fig. 1–6).

Both of these public images — that scientists are more than human, or that they are less than human — are at least somewhat unrealistic. The 1968 publication of *The Double Helix* by Nobel scientist James Watson helped to shatter these misconceptions. Readers of Watson's book discovered that scientists possess many of the same characteristics (both good and bad) as other human beings.

Are scientists today much different from scientists in the past? As noted above, there certainly are many more of them. Research scientists, often very competitive, frequently engage in races to establish the priority of their work by publishing their findings before other scientists. Thus, in addition to an increasing number of scientific publications, we find many more short scientific articles printed in journals that come out very

Figure 1–6 "Dr. Chambers is unscrambling messages from outer space, Dr. Waddell is working on computer language, and Dr. Saville has been conversing with dolphins, but perhaps you could all find some common form of *human* communication." (From Am. Sci. *60*:337, May-June, 1972. Courtesy of Sidney Harris and American Scientist.)

frequently. Scientists often talk of being "scooped"; they fear that others may discover and publish similar findings before them. Indeed, such a thing is not unusual. One study found that among 1400 scientists, two-thirds of them said that they had been "scooped," many of them on several different occasions. Thus, modern science is characterized by an intensely competitive race to be the first to discover significant principles or facts.

Is this intense competition or race to establish priority characteristic of modern science only? Robert Merton suggests that it has been typical of science throughout history. The eighteenth-century writings of Isaac Newton, for example, contain many references to his intent to establish priority over Leibnitz in the invention of the calculus. Modern science, in fact, may be less competitive than science of the past. Elinor Barber found a decline over the years in the frequency with which simultaneous discoveries were the occasion of intense priority conflicts between scientists.

In the early years of science, scientists were concerned with very basic questions regarding the structure and organization of the universe. These questions dealt with man's place in the universe and hinged upon basic philosophical views. Arguments as to whether the earth was the center of the universe or whether the earth revolved around the sun were directly related to basic questions dealing with man's place in relation to nature. In contrast, most scientific questions today are concerned with specific problems; for example, the mode of action of a particular enzyme in converting a specific chemical compound from one form to another. Such detailed questions are not so basic and, therefore, may not lead to such heated debates within the scientific community.

Do the personality traits of scientists differ from those of other people in the population? Anne Roe, in a study of 64 prominent research scientists, found them to possess a better than average amount of two traits — curiosity and intelligence. The median IQ of scientists in her study was 166 (100 = average for population). With the exception of physicists, all of the scientists had been avid readers throughout childhood. Most of the social scientists were active socially from an early age; the other scientists had interpersonal relations of low intensity. They were not talkative, were rather asocial, disliked interpersonal conflict and aggression, and were shy and late-maturing. They demonstrated a much stronger preoccupation with things and ideas than with people. They enjoyed apparent but resolvable disorder and liked to take calculated risks. They had powerful egos and strong control over their impulses, and most were very involved in outside interests. They were independent and not subject to group pressure. They were hard working, enjoyed what they were doing, and were repaid to a large extent by the approval of other scientists.

Professors Garvin McCain and Erwin Segal in their book *The Game of Science* discuss the similarities between basic science and other "games" that people play. The basic scientist plays the "scientific game" for understanding; it is a game, as McCain and Segal say, "whose chief delights are the addition of one neatly contrived stroke that helps give form to a picture; a game affording a glimpse of what no man has seen before; a game from which may come ecstasy by bringing order out of chaos."

Scientists often become extremely enthusiastic and emotionally involved in their work. They enjoy imposing order on chaos. They often compete with other scientists in an attempt to have their ideas accepted by others. Scientific findings are tentative; they must be subjected to validation, testing, and confirmation by other scientists. As the philosopher and mathematician Bertrand Russell has said, "It is not what the man of science believes that distinguishes him but *how* and *why* he believes it. His beliefs are tentative, not dogmatic; they are based on evidence, not on authority."

The scientific subculture is not limited by international boundaries. It is an international undertaking. A scientific breakthrough published in one of the growing number of scientific journals becomes the shared knowledge of the entire world. A scientist requires information, and his whole training and mode of operation work against secrecy in basic science.

John Steinbeck, a biologist by early training, in his book *The Log from the Sea of Cortez*, expressed his view of biological scientists: "Biologists deal with life, teeming boisterous life, and learn something from it—learn that the first rule of life is living."

As in other fields, there exist at present barriers to the entrance or advancement of women in science. Studies by economists G. E. Johnson and F. P. Stafford of Michigan indicate that given equal qualifications and experience, women Ph.D. scientists tend to receive significantly less salary than their male counterparts. The percentage of women taking advanced degrees in science has declined over the past 60 years. Women, especially those who have experienced interrupted or discontinued careers, have difficulty in finding opportunities for acceptance into science. For the years 1964 to 1968, women were awarded less than 0.3 per cent of research grants although they comprised 5 to 8 per cent of scientists. Furthermore, women scientists often receive less salary than men for the same positions of teaching or research. Such barriers and underutilization of potentially qualified women in science represents a great loss to the advancement of science and society.

Research scientists today are employed by government agencies such as the US Department of Agriculture, the National Institutes of Health, and the Environmental Protection Agency, and by universities and private industry. Many scientists are becoming involved in political and social issues. Competent scientific and technological decision-making forms an essential part of our highly technical society, for today there are very few problems facing society that do not involve, in one way or another, science or technology.

Both the role and the resulting impact of science on society have changed drastically in recent years. Prior to the nineteenth century, science existed in a world of values that were opposed to science. In order to survive, science had to avoid conflict with established beliefs and values. Science submerged, ignored, or did not publicize the fact that its findings could lead to a change in human values. Today scientists are considered the new Brahmins, the leaders, who through scientific research and technology set the standards for the world to follow. In this situation, science can no longer claim nonresponsibility for its findings. A growing number of scientists now realize that what they do in the laboratory may have far-reaching effects on the outside world. Not all the consequences of research are foreseeable, but the scientist has the challenging responsibility of ensuring that the public is informed about alternative possibilities or applications of new findings. In the following chapters we will examine these possibilities.

REFERENCES

Basalla, G.: The spread of western science. Science *156*:611–622, 1967.

Blackburn, T. R.: Sensuous-intellectual complementarity in science. Science *172*:1003–1007, 1971.

Bronowski, J.: The Doubleday Pictorial Library of Technology: Man Remakes His World. New York, Doubleday and Co., 1964.

Brooks, H.: Can science survive in the modern age? Science *174*:21–30, 1971.

Burnham, John C.: Science in America: Historical Selections. New York, Holt, Rinehart and Winston, 1971.

Butterfield, H.: The scientific revolution. Sci. Am. *203*:173–192, September 1960.

Butterfield, H.: The Origins of Modern Science. New York, P. F. Collier Inc., 1962 (paperback).

Compton, W. D., ed.: The Interaction of Science and Technology. Urbana, University of Illinois Press, 1969.

*Crowther, J. G.: A Short History of Science. Toronto, Methuen Publications, 1969.

Daniels, G. H.: Science in American Society. New York, Alfred A. Knopf, Inc., 1971.

Darwin, F., ed.: The Autobiography of Charles Darwin and Selected Letters. Reproduced by Dover Publications, New York, 1958.

David, E.: The relation of science and technology. Science *175*(4017):13, 1972.

de Solla Price, Derek J.: Science Since Babylon. New Haven, Yale University Press, 1961.

Eiduson, B. T.: Scientists: Their Psychological World. New York, Basic Books, 1962.

*Recommended further reading

*Fischer, R. B.: Science, Man and Society. Philadelphia, W. B. Saunders Co., 1971.

Furth, R.: The limits of measurement. Sci. Am. *183*:48–51, 1950.

Glass, B.: Science: endless horizons or golden age? Science *171*:23–29, 1971.

Graham, P. A.: Women in academe. Science *169*:1284–1290, 1970.

Jaffe, B.: Crucibles: The Story of Chemistry. Greenwich, Connecticut, Fawcett Publications Inc., 1959.

Klaw, S.: The New Brahmins: Scientific Life in America. New York, William Morrow and Co., Inc., 1968.

Lewin, A. Y., and L. Duchan: Women in academia. Science *173*:892–894, 1971.

Mansfield, E.: Contribution of R & D to economic growth in the United States. Science *175*:477–486, 1972.

*McCain, G., and E. M. Segal: The Game of Science. Belmont, California, Brooks/Cole Publishing Co., 1971.

Merton, R. K.: Behavior patterns of scientists. Am. Sci. *157*:1–23, 1969.

Morison, R. S.: Science and social attitudes. Science *165*:150–156, 1969.

Nasr, S. H.: Science and Civilization in Islam. New York, New American Library, 1968 (paperback).

Parsegian, V. L., et al.: Introduction to Natural Science. Part I: The Physical Sciences. New York, Academic Press, 1968. (Comprehensive text in physical science.)

Roe, Anne: The Making of a Scientist. New York, Dodd, Mead, and Co., 1953.

Roe, Anne: The psychology of the scientist. Science *134*:456–459, 1961.

Rose, H., and S. Rose: Science and Society. Middlesex, England, Penguin Books, 1969 (paperback).

Rossi, A. S.: Women in science: why so few? Science *148*:1196–1202, 1965.

Simpson, G. G.: Biology and the nature of science. Science *139*:81–88, 1963.

Singer, G., et al., eds.: A History of Technology. New York, Oxford University Press, 1954.

Stent, G. S.: The Coming of the Golden Age. Garden City, New York, Natural History Press, 1969.

Toffler, A.: Future Shock. New York, Random House, 1970.

Walsh, J.: The nineteen sixties: a not so fond farewell. Science *166*:1605, 1969.

Weinberg, A. M.: Impact of large-scale science on the United States. Science *134*(3473):161–164, 1961. (Big science has emerged with administrative and budgetary problems.)

White, M. S.: Psychological and social barriers to women in science. Science *170*:413–416, 1970.

---

*Recommended further reading

PROBLEMS

1. Why during history have technological advances often preceded scientific advances?

2. Has technology always benefited society?

3. How will future technology foster further scientific developments?

4. Why do scientists, technologists, and legislators have difficulty in getting together to make improvements (e.g., pollution control)?

5. What are science and technology, and how do they differ from each other?

6. What is meant by "scientific truth"?

7. What role did the early Islamic culture of the Middle East play in the historical development of science?

8. Design a list of questions concerning the nature of scientists themselves (e.g., how did they first become interested in science?), and the role of science and technology in modern society. Interview several scientists from your community, and report to the class. A tape recording, *with the permission of the person interviewed*, may be useful. The class may discuss and compare the scientists interviewed. Do they have similarities as a group? What individual differences are apparent?

BRAINSTORM

1. Is science good, bad, or ethically neutral?
2. Are there limits to what we can learn about nature, or will scientific knowledge continue to grow indefinitely?
3. Do you think scientific progress will eventually cease? If so, why? Will there still be scientists? Students may consider the following: The physical materials of the universe are probably finite (see Chap. 2), but not all entities are finite, e.g., real numbers. Thus, it is at least conceivable that ideas are also not finite. Also, resource depletion, population growth, and environmental pollution may put stresses on advanced technological cultures so that exponential growth will be limited by some environmental factors.

2

THE PHYSICAL UNIVERSE — EXPANDING HORIZONS

You are free before the sun of the day,
and free before the stars of the night;
And you are free when there is no sun
and no moon and no star,
You are even free when you close your
eyes upon all there is.

Kahlil Gibran

For more than a million years, generation after generation of human beings have gazed at the starlit night sky, observed the progression of seasons, watched the sun rise and set as time passed them by, and wondered about their place in nature. What material composes the natural world? What is time? How did the universe begin? Is there intelligence outside our world? Such questions, probing the very nature of existence, comprise a fundamental basis for scientific inquiry. The search for new knowledge is, after all, the search for ourselves and our place in nature.

An important part of the game of science is the development of a limited set of hypotheses or principles that will account for the greatest variety of events. The scientist arrives at seemingly workable principles by induction and then reverses the process, uses the principles to indicate new facts to be observed, and tests to see if the new facts validate the proposed principles. As an example of how scientific explanations evolve, are refined, and repeatedly tested, let us look briefly at the historical development of man's picture of the earth and solar system.

The early Egyptians and Babylonians observed and charted the positions of the stars more than 3000 years ago and noted that several stars appeared near the same place on consecutive days, but over longer periods of time changed their positions quite radically. These "stars" (which we now know as planets) were called *wanderers*. Other early civilizations including the ancient Britons (1800 BC) in Europe, and the Mayas (1000 AD) and Incas (1100 AD) in America made detailed observations of the movements of the stars and planets in order to accurately measure the progression of time and the seasons.

Aristotle (384 to 322 BC) suggested that the earth was a stationary sphere in the center of the universe and that all the "heavenly bodies" were huge transparent spheres that revolved around the earth in perfect circles. The sun, moon, and each planet had its own sphere revolving around the earth and all the fixed stars were on a fixed sphere farther away from the earth than the others. The different motions of the planets were explained by the assumption that different spheres traveled around the earth at different speeds. All the heavenly bodies were perfect and unchanging and each of them traveled along one or more perfect circular paths. Aristotle's theory, however, did not explain why each orbit had to be independently plotted, or why the planets orbit at their particular rates, nor was there any indication of the nature of the huge sphere which housed the stars and was the ceiling of the sky. Other astronomers considered certain modifications, assuming, for example, that the planets did not travel at a constant speed; however, simply stating that the planets changed speed was not a satisfactory explanation. Why did the planets appear to change their speeds?

In the second century AD, Ptolemy of Alexandria modified Aristotle's theory and claimed that the planets traveled in small circular "epicycles" while circling the earth. He was able to predict with limited accuracy where the planets were likely to be at a particular time. His predictions included the assumption that the earth was not always the center of rotation of the planets, but he had no explanation for the eccentricity of the orbits.

In the sixteenth century Nicolaus Copernicus found he could explain the orbital data (a planet's position at different times) by making the assumption that the sun and not the earth was the center of the universe and that the earth revolved (as did the other "wanderers") around the sun. He hypothesized that the earth rotated on its own axis once a day, so that the stars, which appear to rotate, were actually stationary. His explanation, which conflicted with both religious dogma and previous scientific theories, revolutionized man's thought and was a vital part of the Renaissance. Copernicus's explanation, however, was not completely satisfactory. Other scientists, checking the paths of the planets more accurately with special instruments, found that they could not describe the paths of the planets' circular orbits around the sun without substantial error.

About 50 years after Copernicus died, Johannes Kepler (1571–1630) studied the data of the astronomer Tycho Brahe and made observations of his own about the paths of the stars and planets. He concluded that a circular path did not always fit the data; Mars, for example, was not always the same distance from the sun, and placing the center of the circle elsewhere did not solve the problem. He attempted to chart the planets' paths as ovals, but sometimes these paths fell inside and at other times outside the ovals. Finally, he tried approximating the orbit with an ellipse. Assuming the sun as one focus of the ellipse, he could predict the path of Mars or the earth with little error (within the accuracy of his instruments). Kepler's observations led to his three laws of planetary motion: 1) planets travel in elliptical orbits with the sun at one focus; 2) a line connecting the sun with the planets sweeps equal areas in equal times (i.e., the planet travels faster when near the sun and slower when it is farther away) (Fig. 2–1); and 3) the ratio of the cubes of the mean distances of any two planets from the sun equals the ratio of the square of their periodic (orbital) times.

Kepler's theory had much greater impact than previous theories because it provided a systematic relationship to explain why some planets travel faster around the sun than others. However, scientific explanations are never complete; Kepler could not explain *why* planets travel in elliptical orbits, nor did he explain *why* their radius vectors sweep equal areas in equal amounts of time. The basis of Kepler's laws remained to be explained by Isaac Newton.

Isaac Newton (1642–1727), after studying mathematics and

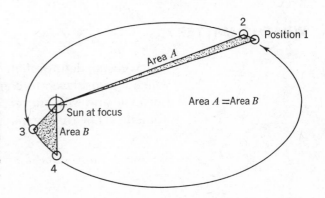

Figure 2–1 The radius vector of Kepler's second law is the line from the sun at the focus to the planet. This line sweeps across equal areas, such as areas A and B, in equal times. Thus, since the distance from position 3 to position 4 is much larger than that from 1 to 2, the planet must travel fastest in its orbit when nearest the sun. (From Parsegian, V. L., et al.: Introduction to Natural Science. Part I: The Physical Sciences. New York, Academic Press, 1968.)

other subjects at Cambridge, began at the age of 24 to consider the problem of planetary motion. Several of his predecessors, including Galileo, had formulated the theory that a body in motion would remain in motion unless acted upon by a force. What was the nature of the force that caused the moon to remain in orbit around the earth instead of going off in a straight line into space? One day, while resting under an apple tree, he was struck on the head by a falling apple. He knew that a body near the earth's surface fell toward the center of the earth with an acceleration of 9.8 meters per sec^2. He asked, "Does the force due to gravity extend out to the moon, and if so, how strong is it at the moon's distance from earth?"

Newton's work, undertaken to solve this question, took 20 years to complete. His book *Principia,* published in 1686, provides the basis of classical physics to the present time. Newton reasoned that a force called *gravity* acts between any pair of bodies that possess mass, and that the magnitude of this force is proportional to the product of these masses and inversely proportional to the square of the distance separating their centers.

$$\text{Gravitational force} = \text{constant} \times \frac{\text{mass}_1 \times \text{mass}_2}{\text{distance }^2} = G \times \frac{m_1 \times m_2}{r^2}$$

This Universal Law of Gravitation in conjunction with other principles formulated by Newton provided the first explanation for such diverse occurrences as falling apples, the movement of the tides, and the movement of the planets described by Kepler.*

The development of the modern concept of the solar system was not, of course, achieved through the insights of the five or six great scientists mentioned earlier. As described in Chapter 1, science proceeds in small steplike progressions, formulating hypotheses, making observations, establishing concepts, testing, and making necessary revisions after further observations.

WHAT IS MATTER?

Webster defines matter as "whatever occupies space; that which is considered to constitute the substance of the physical universe, and with energy, to form the basis of objective phenomena." The theory that matter is composed of discrete par-

---

*For a detailed explanation of Newton's work see Parsegian et al., 1968, pp. 125–151.

ticles, or atoms, can be traced as far back as the ancient Greeks, but during the present century, concepts of matter and atomic structure have undergone profound and radical changes. What is the contemporary physicist's view of atomic structure and how does it differ from earlier models? To answer this question we shall look at the development of modern physics during the last fifty years.

Early in this century, experiments performed by Ernest Rutherford led to the proposal that the center of the atom is occupied by a very tiny, dense, positively charged nucleus. The remainder of the atomic volume is occupied by minute, energetic, negatively charged electrons. J. J. Thomson, Robert A. Millikan and others at the turn of the century first characterized the electron and showed that it had a mass only 1/2000 as large as the lightest atom (hydrogen). Rutherford's experiments resulted in a model in which almost the whole mass of the atom is associated with the nucleus, which occupies only *one millionth of a millionth* of the atomic volume. Hence the atom must contain a great deal of empty space.

ELECTROMAGNETIC RADIATION

Different types of energy, from gamma radiation to light and radio waves, can be considered electromagnetic radiation. In electromagnetic waves, an oscillation of an electric field is coupled at right angles to the oscillation of a magnetic field and both of these oscillations are at right angles to the direction of propagation of the energy. The wave traces out a sine wave function that propagates outward at a speed of 3×10^8 m/sec $= c$ (Fig. 2–2A).

For example, in a radio wave, a transmitting aerial (a long metal rod or wire) carries an electric current (a flow of free electrons) that typically changes its direction of flow a million times per second. This oscillation produces a disturbance which travels outward at the speed c. The number of wave crests passing a given point in space each second is called the frequency and is expressed as cycles per second (cps) or *hertz* (Hz). The frequency (ν) of electromagnetic radiation is related to the wavelength as $\nu = \dfrac{c}{\lambda}$, where $c = 3 \times 10^8$ m/sec and $\lambda = $ the wavelength in meters.

All types of electromagnetic radiation travel at speed c; they differ only in their wavelength, and hence, frequency. Thus, long-wavelength radio waves have a wavelength of 10^7 meters and a frequency of 30 hertz; heat is a form of electromagnetic energy in the infrared range of wavelengths from about 10^{-4} to 10^{-5} meters. Visible light includes a narrow band of wavelengths between 7×10^{-7} and 4×10^{-7} meters. Ultraviolet and x-rays are of shorter wavelength, and gamma rays are of extremely short wavelength and high frequency (Fig. 2–2B).

Because electromagnetic waves travel at a finite speed c, they take some time to cross long distances. For example, electromagnetic waves (light or radio waves) would take about 26 minutes to make the round trip between earth and a

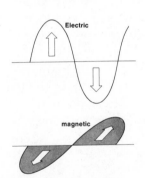

Electric

magnetic

Figure 2–2 *A, B,* and *C,* In an electromagnetic wave, the electric and magnetic fields are in phase perpendicular to each other and at a right angle to the direction of propagation. (Legend continued on following page.)

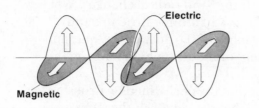

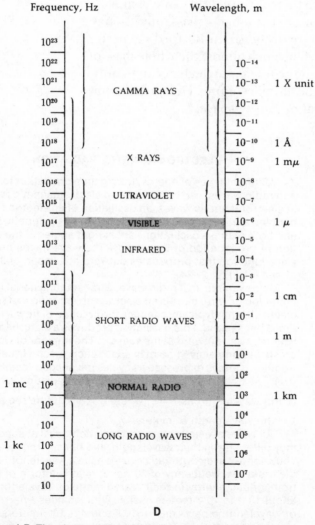

D

Figure 2-2 *D* and *E*, The electromagnetic spectrum ranges from extremely short wavelength and high frequency gamma rays or x-rays up to long wavelength–low frequency radio waves. About one million x-ray wavelengths could fit within the diameter of a needle. (Fig. 2–2 *D* from Bueche, F.: Physical Science. N.Y., Worth Publishers Inc., 1972, p. 191.)

spaceship near Mars. A *light-year* is the distance traveled by an electromagnetic wave (light) traveling at speed *c* in one year, that is, about 10^{16} meters.

Electromagnetic radiation exhibits a duality. It can be considered as *either* a wave or a *particle* phenomenon. In 1900, the German physicist Max Planck introduced the concept that radiation can exist as discrete quantities called *quanta* of energy and not at intermediate values:

$$\Delta E = h\nu$$

where ΔE = the amount of energy in an energy change,
ν = the frequency of the oscillator,
and h = Planck's constant (6.625×10^{-27} erg · sec)

Thus, energy increases at higher frequencies (shorter wavelengths). A quantum (hν) of blue light contains more energy than a quantum of red light; and gamma rays of extremely high energy are called ionizing radiation because they are of high enough energy to remove electrons from atoms and molecules with which they collide, causing formation of positive ions. Ionizing radiation is also capable of penetrating living tissue with resultant damage to cells (see box insert, Chap. 7).

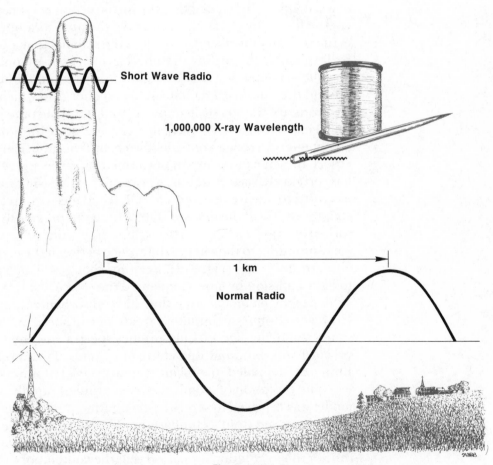

Short Wave Radio

1,000,000 X-ray Wavelength

1 km

Normal Radio

Figure 2–2   E.

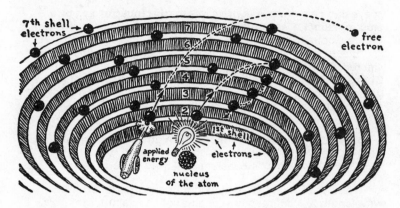

Figure 2–3 Schematic view of Bohr model of the atom with surrounding electron shells. The outer shells are of higher energy. (From Murchie, G.: Music of the Spheres. New York, Dover Publications, Inc., 1967.)

In 1911, a young Danish physicist, Niels Bohr, joined Rutherford at the Cavendish Laboratory of Cambridge University in England. At this time the atom was believed to be similar to a tiny solar system, with the nucleus surrounded by revolving electrons. However, according to the laws of classical physics, such an atomic model would lose energy continuously, with the electrons collapsing into the nucleus in one hundred-millionth of a second. Bohr argued that the introduction of "quantized" energy levels for electrons would save the nuclear atom from extinction and satisfactorily explain the stability that atoms in reality exhibit. According to Bohr, electrons possess only certain allowed energy states corresponding to certain orbits around the nucleus. The state of lowest energy in which the electrons are found is called the ground state; the others are called excited states. To move to an excited state the atom must accept energy from an outside source, but the atom can only absorb this energy in certain prescribed capsules. Every atom has certain characteristic energy capsules which will cause the electrons to move to excited states. Excited electrons return spontaneously to the ground state, losing the previously absorbed energy as a light particle (photon) having a frequency corresponding to the energy difference between the excited and ground states (Fig. 2–3). Relating changes in energy states to light emission by atoms explained the characteristic spectra exhibited by elements after they have absorbed energy. However, Bohr's quantitative treatment of electron energy states could not successfully predict spectra for any elements except hydrogen, and one-electron systems. Furthermore, Bohr's model failed to explain most aspects of atomic behavior.

The evolution of atomic theory continued rapidly. Bohr's model was replaced by a relativistic treatment of the electron which removed Bohr's emphasis on the electron as a particle and substituted the concept of the electron as a wave. It was during this revolutionary period that the contemporary picture of atomic structure emerged. Louis V. de Broglie began in 1920 to study the motion of electrons in atoms. It was known that light and other electromagnetic radiation could behave as either a wave or a particle. According to de Broglie, an electron, like a photon, has a wave motion with a wavelength corresponding to its velocity (Fig. 2–4). Because these electron waves are confined by attraction to the nucleus, the question is what effect this confinement will have on the waves. In 1926, Erwin Schroedinger, an Austrian at the University of Zurich, proposed that the quantum jumps of an electron to another energy state as proposed by Bohr really represented an electron moving from a state that could accommodate a whole (or integer) number of wavelengths associated with its velocity to another orbit, which could again accommodate a whole number of its new wavelengths corresponding to its new velocity. For his work Schroedinger earned the Nobel prize in physics (1933).

In 1925, Werner Heisenberg proposed the *uncertainty principle,* stating that it is inherently impossible to determine simultaneously the location and the velocity of an atomic particle. At this time, another group of scientists including Max Born suggested that the wave representations of de Broglie and Schroedinger could not be interpreted as being waves in the usual sense. They proposed instead that one can really only determine the *probable* location of an electron at any time. Thus, we no longer speak of circular orbits of electrons but, simply, of the area of high probability for electron location. Electrons are thought to represent a sort of energy cloud surrounding the nucleus. In recent years, high-speed digital computers have been used to analyze details of atomic and subatomic structure and chemical bonding (see Chap. 8).

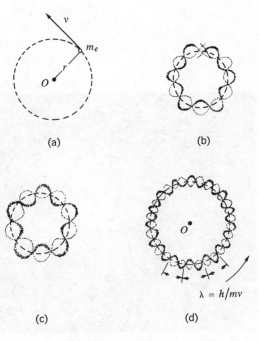

(a)

(b)

(c)

(d)

Figure 2–4 Comparison of the de Broglie wave model with the earlier Bohr model. (a) Electron circular orbit as envisaged by original Bohr model. (b) According to de Broglie, the electron has a wave motion, with a wavelength (λ) corresponding to $\lambda = h/mv$ where m = mass, v = velocity, and h = Planck's constant. In this drawing the six waves do not quite fit into this circumference; therefore this orbit is not possible to the electron. (c) The five wavelengths meet exactly ro form standing waves. This orbit is possible. (d) This orbit accommodates 16 wavelengths exactly. The waves can propagate in either sense, but they must reinforce each other on successive rotations. (From Parsegian, V. L., et al.: Introduction to Natural Science. Part I: The Physical Sciences. New York, Academic Press, 1968.)

WHAT IS ANTIMATTER?

We have briefly reviewed the development of the contemporary view of matter, and the concept of an atom containing a positively charged nucleus surrounded by a cloud of negatively charged electrons. But could an atom exist that was built on the opposite structure, with a negatively charged nucleus surrounded by positively charged particles? P. A. M. Dirac predicted mathematically that an electron could indeed have an opposite with exactly the same mass, and in 1932 the positron (positively charged electron) was detected and identified in the laboratory. Ever since the discovery of the positron, physicists have wondered about the possibility of antimatter. If an antielectron could exist, why not an antiproton, or an antineutron? In recent years, all of these particles have been created in the laboratory. If antiparticles exist, are there anti-atoms or antiworlds or even antilife?

Can antiparticles actually exist in the universe apart from their artificial production in laboratory particle accelerators? Even if antimatter exists in outer space, it could not be seen with a telescope, because antimatter would look exactly like ordinary matter. However, the question is not insoluble. When antimatter and matter collide, the mass of both particles is converted into energy in the form of radiation. Some of this radiation takes the form of radio waves. We know that radio waves originate from some points in space even within our own galaxy. It is possible, for example, that the radio emissions of the Crab Nebula arise from electrons and positrons which have been created by the collision of matter with antimatter (Fig. 2–5). Some astronomers believe that Cygnus A represents a pair of galaxies in collision, and astronomers Burbridge and Hoyle

Figure 2–5   The Crab Nebula. The remnants of a cosmic explosion, equivalent in violence to the simultaneous explosion of about 1,000,000,000,000,000,000,000,000 hydrogen bombs. This cloud now has a diameter of some thirty million million miles. (From Hoyle, F.: The Nature of the Universe. New York, Harper & Row, 1966. Reproduced with permission from the Hale Observatories.)

have estimated that the annihilation of antimatter in such a colliding galaxy would generate a total of about 2.3×10^{36} calories per second of energy. According to the measurements of radio astronomers, Cygnus A is actually emitting radiation energy at this rate.

If matter and antimatter exist in our universe, how are they kept from converging and destroying each other? One explanation might be antigravity. Experiments designed to demonstrate the existence of antigravity, according to Burbidge and Hoyle, may be possible.

WHAT IS THE NATURE OF TIME AND SPACE?

We have discussed briefly the concepts of matter and antimatter, but what do we mean by *space* and *time* when we describe the universe? During the nineteenth century, scientists realized that what we call light can actually be considered electromagnetic energy of different wavelengths traveling through space. Waves such as sound, however, require some medium for their transport. Sound can travel through air or through water but not through a vacuum. Since outer space is a near-vacuum, how do light rays travel from the sun or from distant stars to the earth?

In order to explain the propagation of light waves through space, scientists in the nineteenth century proposed the existence of a "luminiferous ether." This ether supposedly permeated the entire universe, and attempts were made to describe its physical properties. A famous well controlled and precise experiment was carried out in 1881 by the American scientists Michelson and Morley for the purpose of demonstrating the existence and physical properties of the ether.* Their carefully designed experiment, however, failed to show the occurrence of the ether. Because much of physical theory at that time depended upon the assumed existence of this luminiferous ether, physicists at the end of the nineteenth century were faced with a crisis. Did the ether exist or not? If it did, why couldn't they detect it?

At this time of crisis in scientific thinking, a minor consultant at the Swiss patent office published theories which eventually revolutionized scientific thinking and contributed to a great leap forward in our understanding of the universe. His name was Albert Einstein. In addition to his famous theory of relativity, he provided the first mathematical analysis of Brownian movement, which gave support to the atomic theory. He also

---

*For an intriguing description of the Michelson-Morley experiment see Coleman (1969) or Parsegian (1968).

introduced the idea that light could be considered a particle or photon and received the Nobel prize for his work on the photo-electric effect. Einstein published the *Special Theory of Relativity* in 1905 and the *General Theory* in 1916. Although the derivations of his theories are mathematically complex, the underlying ideas are not difficult to understand (see box insert).

RELATIVITY PHYSICS

According to Einstein's special theory of relativity, the speed of light *(c)* is a universal constant and remains the same (3×10^8 m/sec) regardless of the position or motion of the observer.

When observing an object traveling at high speeds (v), approaching the speed of light (c), an outside observer could except rather surprising changes to occur:

1. *Length.* At high speeds an object contracts in length (becomes shorter) according to the relationship

$$L' = L \sqrt{1 - \frac{v^2}{c^2}}$$

where L' = observed length of traveling body
L = original length of body at rest

2. *Mass.* At high velocities an object increases in mass

$$m' = \frac{m}{\sqrt{1 - \frac{v^2}{c^2}}}$$

where m' = observed mass of traveling body
m = original mass of body at rest

Thus, for example, a 135 kilogram man running at 24 kilometers per hour will increase in mass about 30 millionths of a gram.

This relationship does not apply to photons or electromagnetic radiation, which according to the theory have a rest mass of 0 (i.e., if they were not moving, they would not exist).

3. *Velocity.* At high velocity, two bodies' speeds relative to each other are no longer additive.

$$V_{AB} = \frac{V_A + V_B}{1 + \frac{V_A V_B}{c^2}}$$

where V_A = velocity of body A
V_B = velocity of body B
V_{AB} = velocity of A with respect to B

Thus, for example, if two bodies are traveling toward each other, each with a velocity of 100,000 km per second, relative to a third observer their velocity relative to each other is not 200,000 km per second (as predicted by classical physics) but rather 180,000 km per second.

According to Einstein's theory, nothing can travel as fast as the speed of light (c), because its length would shrink to nothing and its mass would become infinite.

4. *Time.* At high speed, time (considered the fourth dimension) slows down relative to an outside observer.

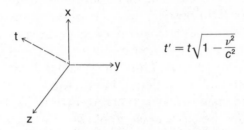

$$t' = t\sqrt{1 - \frac{v^2}{c^2}}$$

where t' = time in moving system according to stationary observer.

t = time according to stationary observer.

Time is "relative": two events may be simultaneous for one observer, but not for all observers.

5. *Energy.* Mass and energy are equivalent and interchangeable according to the relationship

$$E = mc^2$$

where E = energy
m = mass
c = speed of light

This last equation is of more than theoretical interest and has found applications of far-reaching significance— the explosive discharge of mass to energy in atomic (fission) and hydrogen (fusion) bombs, as well as the useful controlled fission heat generation used to drive steam turbines to generate electrical power.

Einstein's General Theory of Relativity.   The special theory of relativity deals with objects in an unaccelerated state outside the influence of a magnetic field. Einstein later expanded this theory into the *general theory of relativity* to include objects undergoing acceleration or under the influence of a gravitational field.

Let us examine the basic concepts of Einstein's theories, starting with the *Special Theory,* which deals with unaccelerating bodies. First, Einstein stated that if the ether exists, it cannot be detected. If it permeates all of the universe, there is no way to demonstrate its existence, because it could be found only with reference to some other point (of non-ether). For example, if you travel far out in space and use no stars or planets as reference points, you will never know whether you are moving or not: thus, Einstein recognized that *all motion is relative.* We can speak of the motion of a body only with respect to something else. Because the whole universe is filled with motion, we cannot say what is moving and what is stationary; we can only say that some bodies are moving relative to others. Thus, we can talk only of *relative motion.*

Nevertheless, as Einstein proposed, there does exist one type of particle whose velocity does not depend upon the frame of reference or the position of the observer. This is the photon, or light particle, which travels at a speed of 3×10^8 meters per second, relative to an observer, regardless of whether the observer is moving rapidly toward the source of light or rapidly away from it. Thus, the speed of light (c) is a universal constant, and does not depend upon the location or the relative movement of the observer. According to classical (Newtonian) physics, the *relative velocity* of two moving bodies is additive. For example, driving down the highway in your car at 100 km per hour, relative to the earth, you will approach a parked car at 100 km per hour. If instead of being parked, the other car is approaching you at 50 km per hour, your relative velocity with respect to each other is 100 plus 50 or 150 km per hour, but according to relativity this is not the case. Consider two rocket ships approaching one another in outer space, each traveling at 100,000 km per second relative to an outside observer. Their relative velocity with respect to each other is not 200,000 km per second, but only 180,000 km per second (see Box Insert, Equation 3). In fact, one of the basic tenets of relativity is that nothing can exceed 3×10^8 miles per second in velocity (the speed of light).

As a body increases in velocity, a contraction in its length and an increase in its mass will be apparent to an outside observer. As it approaches the speed of light, the mass increases without limit. This contraction in length and increase in mass is apparent only to an outside observer. A spaceman traveling in a rocket at 100,000 km per second would not be aware of, nor be able to detect, any change in the length or mass of his own body or of the spaceship.

All electromagnetic waves, including light, travel at 3×10^8 meters per second in a vacuum. Is their mass, therefore, infinite? The answer is no. These relationships simply do not apply to light waves or photons. The mass of a photon at rest is equal to zero; a photon cannot exist when it is not traveling at the speed of light.

Einstein made the important generalization that mass and energy are equivalent; that is, $E = mc^2$ where E is the equivalent energy, m the mass of the body, and c the velocity of light. His now famous insight has had far-reaching practical consequences in the development of atomic power (see Chaps. 4 and 5).

WHAT IS TIME?

Before Einstein, time was thought to flow in one direction, like a river flowing uniformly down a mountain, passing all bystanders at the same rate. The special theory of relativity, however, asserts that time flows at *different* rates for two observers moving relative to each other. Consider how you might determine the dimensions of the room in which you are now sitting. You would measure first the length, then the width, and then the height of the room. You would thus have a description of the room in three dimensions. Similarly, in classical terms we define the direction of travel of an object such as an airplane with reference to some coordinate, using X, Y, and Z axes to locate its position in three-dimensional space. According to relativity, however, for a complete description we must also take into account its position in relation to a fourth dimension— time.

The general relativity theory also tells us that as a system accelerates and its speed increases, time (according to an outside observer) slows down. Let us consider an example of this *time dilation.* Imagine a pair of twin brothers, one named Earthman and the other Spaceman. Spaceman decides to take a trip through space to Arcturus, a star which is 33 light-years from the earth. If his spaceship travels at a velocity close to that of light, he will arrive at Arcturus slightly more than 33 years later, earth time. If he returns immediately, he will arrive back on earth approximately 66 years after leaving the earth. Since he is traveling at high velocity, time has slowed down and to Spaceman only one day has elapsed from the time he left until he arrived back on eath. However, he will find his brother, Earthman, 66 years older and perhaps even passed away and gone.

The *general relativity theory* predicts that a moving clock will record less time than a clock at rest. This relativistic "clock paradox" has been the subject of extensive scientific debate during the twentieth century. The recent development of atomic clocks has allowed scientists to measure time with extreme accuracy and confirm the predictions of relativity theory. Such clocks work according to the frequency of the cesium atom, which, when electrically excited, emits light at a frequency of precisely 9,192,631,770 cycles per second. Scientists armed with this new and highly accurate standard of time realized that an experimental test of Einstein's clock paradox was now possible.

During October 1971, four atomic clocks were flown on regularly scheduled jet flights around the world, once eastward and once westward, to test Einstein's theory of relativity. According to theory, flying clocks compared to stationary reference clocks should have lost time (aged slower) during the eastward trip, and gained time (aged faster) during the westward trip. The experimenters found that relative to the atomic time scale of the United States Naval Observatory stationary clock, the flying clocks lost 59 nanoseconds (billionths of a second) during the eastward trip and gained 273 nanoseconds during the westward trip. This was in agreement with the predicted results, and they concluded that "there seems to be little basis for further arguments about whether clocks will indicate the same time after a round trip, for we find that they do not."

Is time like a river flowing in only one direction or can it flow in more than one direction? Our everyday intuitive experience teaches us that time is irreversible. We are born, grow older and die. In the nineteenth century physicists proposed that the *direction* of time was actually related to the increasing entropy of our universe. Entropy can be defined as increasing disorder. Sir Arthur Eddington stated the concept in this way:

Let us draw an arrow arbitrarily. If, as we follow the arrow, we find more and more of the random element in the state of the world, then the arrow is pointing towards the future; if the random element decreases the arrow points toward the past. That is the only distinction known to physics.

Although the universe as we know it is increasing in entropy and expanding, recent theorists have speculated that there may be pockets of negative entropy within our universe where order is increasing.

Time is an elementary parameter serving to identify the order or sequence of events. To give quantitative meaning to events, man constructed calendars, and divided time into established and continuing sequences of recurring events (e.g., the revolution of the earth around the sun). The point at which the calendar begins is arbitrary, as is the choice of associating BC with negative time and AD with positive time. It depends on the choice of coordinates.

Space (usually described in three coordinates) can be reversed. The choice of right- or left-handed coordinate systems is simply a matter of convention. Such reversed coordinates can be depicted as mirror images of each other (Fig. 2–6). Recent evidence now suggests that time as well as space may be reversible.

The universe, the most complex system known, is thought to have started from an ordered condition—perhaps a collection of protons distributed uniformly throughout space or possibly distributed uniformly on the surface of a sphere. The evolutionary process of the universe is thought to have proceeded with time toward a disordered state. This evolution from gravitational collapse to the generation of elements,

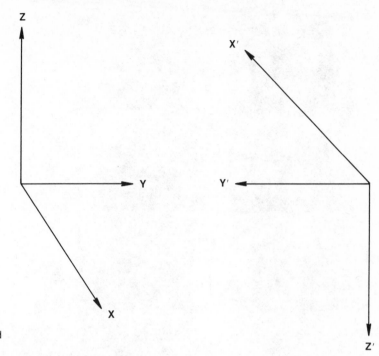

Figure 2–6 Right- and left-handed
coordinate systems.

through nuclear reactions and the formation of chemical compounds, to the creation of living organisms and their evolution,
if viewed under a different coordinate system, could be seen as
a motion system of the evolutionary process in reverse. Because
of the large number of variables in a system such as the universe we cannot conceive that the reversed evolution from the
disordered to the ordered state could ever take place. This
gives us our sense of absoluteness of the direction of flow of
time from an ordered to a disordered state that we call increasing entropy. Large systems such as human actions or the evolution of the universe are not suitable for tests of time reversal
because the initial conditions play an overwhelming role in determining the motions of the system, but it is possible to test the
principle by studying subatomic particles and their interactions.

Atom smashers, used in the study of subnuclear particles,
accelerate particles (usually protons) to extremely high velocities and then hurl them into stationary targets. On impact the
nuclei of the target atoms break apart, scattering fragments
(subatomic particles). In the new 500-billion electron volt (BeV)
accelerator recently constructed at Geneva, Switzerland, particles are sent alternately in clockwise and counterclockwise directions and beamed precisely so that they collide with one
another. With this new type of accelerator scientists hope to
produce energies of over 100,000 BeV. Such fantastic power
can be used to bring together protons and their antimatter op-
posites, antiprotons, and may provide new answers concerning
the nature of time, matter, and the universe itself.

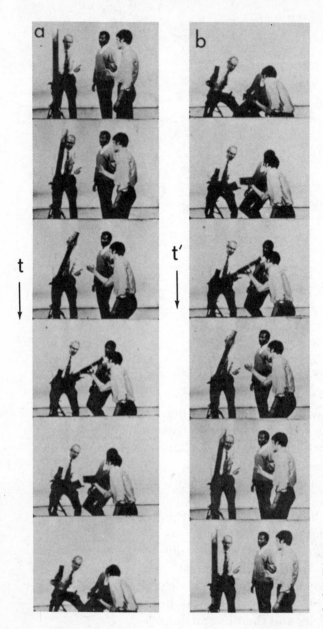

t ↓

t' ↓

Figure 2–7 Time reversal in a motion picture sequence. (From Sachs, R. G.: Time reversal. Science *176*:587–597, 1972. Copyright 1972 by the American Association for the Advancement of Science.)

In 1948, Richard P. Feynman, who shared the 1965 Nobel prize in physics, developed a theory in which an antiparticle is regarded as a particle moving backward in time. Thus, a positron is simply an electron moving momentarily backward in time. If this is the case, then antimatter may represent matter moving backward in time with respect to our observations. In fact, there are mathematical and physical data dealing with the symmetry of elementary particle interactions and the universe as a whole which indicate that time indeed can run backwards! Instead of the mirror used for space reversal, we may imagine time reversal as a motion picture run backwards (Fig. 2–7).

HOW DID THE UNIVERSE BEGIN—OR DID IT?

Let us briefly consider what Shklovskii and Sagan have called "the most momentous question of contemporary natural science—the cosmological problem." *Cosmology* is the study of the structure and development of the entire universe. Is the universe finite in space and time or is it infinite? Does it have physical boundaries? Does it have a beginning or an end? The earth, revolving around a star called the sun, is located about two-thirds of the way out toward the edge of a large galaxy. The galaxy, the Milky Way, may contain in excess of 100 billion stars. The nearest star, *Alpha Centauri,* is 4.3 light-years away, that is, if we traveled at the speed of light, it would take us 4.3 years to reach this star. Our neighboring galaxy, *Andromeda,* lies 700,000 light-years distant. There are one billion other galaxies in the range made visible by use of large light telescopes (Figs. 2–8 and 2–9).

Until recent years the observational evidence from cosmol-

Figure 2–8 The great Andromeda galaxy is a giant star system similar in shape and structure to our own Milky way. Although it can be seen with the naked eye as a faint luminescence in the constellation Andromeda, it is 700,000 light years away. Yet it is the nearest of all the island universes that wheel in the depths of space. Its diameter is 60,000 light years. To an observer situated in this galaxy, our Milky Way would look very much like this. The smaller nebulosities nearby are minor members of the super-galactic cluster that encompases the Andromeda spiral, our Milky Way, and the Magellanic Clouds. (From Barnett, L.: The Universe and Dr. Einstein. New York, The New American Library, 1960. Reproduced with permission from the Hale Observatories.)

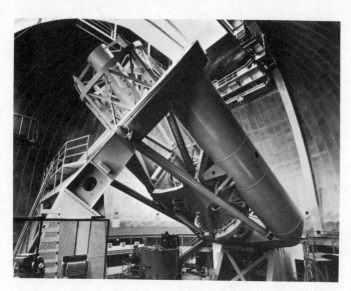

Figure 2–9 The 500 cm Hale Telescope. Notice the flight of steps and the door at bottom-center. These fix the huge scale of the instrument. (From Hoyle, F.: The Nature of the Universe. New York, Harper & Row, 1966. Reproduced with permission from the Hale Observatories.)

ogy was so scarce that it could hardly be considered a science. However, new knowledge, much of it derived from radio astronomy, has made cosmology a much more empirical subject. Radiotelescopes extend observations into regions of the spectrum never before observable (including x-rays, gamma-rays, and infrared); they also greatly extend the boundaries of the observable universe (Fig. 2–10). Starlike objects called *quasistellar radio sources (quasars)* can now be identified from as much as 10 billion light-years away.

In recent years evidence has been mounting that one source of x-rays in our galaxy is a *black hole*—a region left behind after a very massive star has collapsed, where the gravitational forces are so strong that nothing—not even light—can escape. Several astronomers have found evidence for a black hole in the constellation Cygnus, and observations are continuing in an attempt to verify this x-ray source as the first identification of the predicted black hole.

Why Is the Night Sky Dark? If the universe is infinite, and galaxies are distributed uniformly throughout space, each emitting light, as we look into the sky our view will include a larger and larger area at greater distances. Thus, the night sky should include an infinite number of stars and should appear solidly bright in background, but this is not the case, and the apparent contradiction is known as *Olber's paradox.* Explanations to resolve this paradox are based on the theory of a *finite* universe. For example, if the universe is finite in dimension then at some far distance, stars no longer exist. Or, as we shall see, if the universe is finite in *time,* that is, if it has a beginning, then we can also expect that at some distance from the origin of an expanding universe, there will be no more stars.

Considerable evidence indicates that the universe is expanding and that the galaxies farthest away from us are receding from us at the fastest rate.* In 1927, G. E. Lemaître proposed a theory that the universe originated from a tremendous primordial explosion. George Gamow later gave widespread support to this *"big bang" theory.* We can extrapolate backward in time to a point at which all matter in the expanding universe was compressed into one densely packed and extremely small mass of neutrons with a density greater than that of an atomic nucleus (10^{15} gm/cm^3). It is estimated that approximately 12 billion years ago, this extremely dense mass of compressed neutrons underwent the greatest explosion of all time, hurtling matter outward and forming our present galaxies, which have continued their outward ex-

---

*For a description of how distances to stars are measured and the velocities of distant stars computed from their "red shift" see Parsegian et al., 1968, pp. 252–260.

Figure 2–10   The Arecibo Ionospheric Observatory, the world's largest radio telescope now in operation. It has an aperture of 300 meters and is semi-steerable. The cables comprising the antenna itself are layered down into a deep depression smoothed out into an already existing valley in Arecibo, Puerto Rico. (From Shklovskii, I. S., and C. Sagan: Intelligent Life in the Universe. San Francisco, Holden-Day, Inc., 1966.)

pansion ever since the original explosion. Instead of originating at a finite time, however, some scientists have proposed that a cyclic expansion and contraction of matter in the universe occurs extending infinitely in both directions of time, having no beginning and no end, but only a cyclic explosion-expansion-contraction-explosion-forever (Fig. 2–11). The big bang theory, however, could not account for the evolutionary formation of the heavier atomic elements in the universe from the original neutrons in the dense mass. Therefore, another theory was proposed.

The steady state or *continuous creation model* of the universe has been advanced by the British astronomer Fred Hoyle and others. According to this theory, the universe is infinitely old. Matter (in the form of hydrogen) is being created (somehow) at exactly the rate needed to fill the void left by the expansion of matter in the universe. This model of the universe takes into account observations indicating that galaxies appear to recede irrespective of the location of an observer. The steady state model considers the universe infinite, yet Einstein assumed that the universe is finite in volume, and that because of the curvature of space, light would eventually travel in a circle. Is it possible to have an infinite steady state universe within a finite space? The answer is yes. Galaxies far from the observer are receding so fast that due to relativity, we find "length contraction" and they become infinitely small and crowded near the edge of the universe.

There are, however, problems with the steady state model. In 1965, Allen Sandage announced the discovery of *blue stellar objects* or BSOs (see Asimov, 1966). Sandage's new evidence supports the big bang theory; in fact, Sandage says that the universe is pulsating (expanding and contracting) with a cycle of about 82 billion years.

In 1965 workers at the Bell Laboratories in New Jersey found remnant microwave background radiation in distant parts of the universe. This evidence lent further support to the big bang theory, and in 1965, even Hoyle retreated somewhat and proposed that we may actually be in a localized big-bang section of the universe.

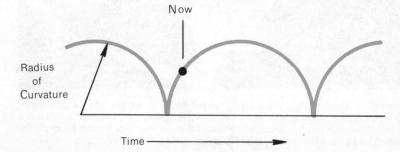

Figure 2–11 Schematic representation of the change of the radius of curvature of a closed, pulsating universe with time. (From Shklovskii, I. S., and C. Sagan: Intelligent Life in the Universe. San Francisco, Holden-Day, Inc., 1966.)

The Swedish scientist Oskar Klein has recently suggested another cosmological theory based on the existence of antimatter. His theory states that the initial state of the universe consisted of an extremely dilute cloud of gas. This cloud, containing both particles and antiparticles, began to contract owing to the mutual gravitational attraction of the particles. When the radius of the cloud shrank to about one billion light-years, the radiation arising from the particle-antiparticle annihilation was so strong that the cloud began to expand again. Matter and antimatter cannot exist together, and Klein's is one of several theories that explain how they maintain their existence within separated pockets of the universe.

Are there other galaxies in the universe composed of antimatter and inhabited by antibeings? If so, their activities would appear to us to run in a time-reversed series, and our activities would appear to them to run in reverse. However, it may be impossible even to observe such a time-reversed galaxy. Light, instead of radiating from it, would seem to be going toward it, and galaxies of matter and antimatter would be totally invisible to each other. Such questions, Martin Gardner states,

> . . . go far beyond the reach of physics and probe aspects of existence that we are as little capable of comprehending as the fish in the river Liffey are of comprehending the city of Dublin.

Rapid advances in both manned and unmanned space flight within the last two decades have provided new and sometimes revolutionary information about our solar system and the universe. For example, the Mariner 9 unmanned space probe, which arrived on a circling orbit of Mars in November 1972, led to major revisions concerning our knowledge of the mysterious "red planet." The detailed photographs and other information gathered by the probe indicate that contrary to previous concepts, Mars is still undergoing sharp climatic changes, large-scale geological activity, and may have enough water on its surface to support primitive forms of life. The photographs relayed back to earth showed large canyons evidently carved out within recent geological time by some flowing liquid (water?).

ARE WE ALONE IN THE UNIVERSE?

Does life exist elsewhere in our solar system or in the universe? Is extraterrestrial intelligent life perhaps attempting to contact us even now? One of the greatest arguments in favor of the presence of life elsewhere in the universe is the basic similarity that science has found in the chemical and physical processes throughout the universe. Most scientists today agree that given the proper conditions, the creation of life in the universe is as natural a process as the formation of the higher chemical elements, the stars, and galaxies.

The idea of life existing beyond the earth is not new. Giordano Bruno, a disciple of Copernicus, stated in the sixteenth century, "Innumerable suns exist . . . innumerable earths revolve about these suns in a manner similar to the ways the seven" (then known) "planets revolve around our sun. Living beings inhabit these worlds." For such heretical views, Bruno was burned at the stake in 1600.

There is evidence that the basic building blocks of life exist elsewhere in the universe. Such organic compounds as ammonia, carbon monoxide, formaldehyde, ethyl alcohol, and water have been identified outside our solar system. Furthermore, laboratory experiments indicate that under the right conditions, such chemicals can combine to form the basic building blocks of life itself. The fact that life on earth in one form or another can withstand extreme environments including ice-bound Antarctica, near-boiling hot springs, strong acids, deep sea trenches, and deserts, is an indication of the great adaptability of life forms. Although simple life forms such as bacteria and algae can survive under such conditions, most exobiologists agree that the likelihood of more advanced, intelligent beings existing on other planets within our solar system is negligible.

But what about other planets revolving around other stars? One nearby star (Barnard's star) shows irregularities in its orbit which suggest to astronomers that two planets about the size of Jupiter and Saturn are orbiting around the star and affecting its gravitational field. Carl Sagan has stated that there are about 10^{11} stars in our galaxy and 10^9 other galaxies visible through our large telescopes. Thus, there exist at least 10^{20} stars in the observable universe, many of them accompanied by planetary systems. Furthermore, estimates suggest that approximately one million solar systems may be formed in the universe each hour.

Not all of the planets surrounding other stars could support life. Among other things, they must be far enough from the star not to be exceedingly hot, and close enough to the star not to be exceedingly cold. In September 1971, Russian, American, and other scientists meeting in the Soviet Union estimated that there are now 100,000 to 1,000,000 technical civilizations within the Milky Way (Fig. 2–12). This number could vary widely depending upon the length of time that such a civilization survived as a technological society. A Rand Corporation study by Steven Dole estimated that our galaxy contains 600 million inhabited planets, or on the average, about 1 inhabited planet within 27 light-years of the earth, 5 within 47 light-years,

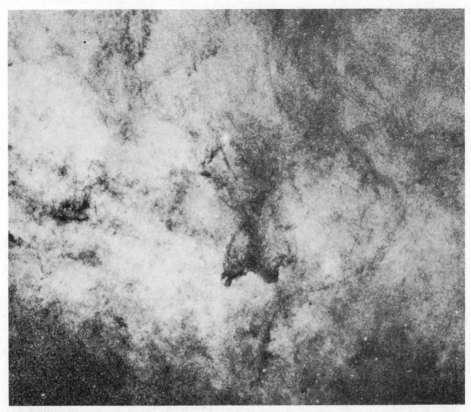

Figure 2-12 A star cloud in the region of the galactic center. There are approximately a million stars in this photograph. According to estimates, a planet of one of these stars holds a technical civilization vastly in advance of our own. (From Shkovskii, I. S., and C. Sagan, Intelligent Life in the Universe. San Francisco, Holden-Day, Inc., 1966. Reproduced with permission from the Hale Observatories.)

10 within 59 light-years, and 50 within 100 light-years—all close enough for us to be able to signal them using present radio technology. Walter Sullivan has examined the question of what would happen if indeed we did make contact with another civilization. Such an extraterrestrial civilization might be far superior and far more advanced than our own (Figs. 2–13 and 2–14).

Scientists are now attempting to contact extraterrestrial civilizations. They are listening with huge radiotelescopes beamed towards distant stars in an attempt to detect messages sent by intelligent beings from other planets.

In 1972 the Pioneer 10 spacecraft was launched from the United States. Its flight will take it to the vicinity of Jupiter, then past that planet, to be accelerated out beyond our solar system. Given the very small probability that an advanced civilization might detect and intercept it, scientists included on the spacecraft a message from Earth. The message, in the form of an engraved plate, carries the representation of a

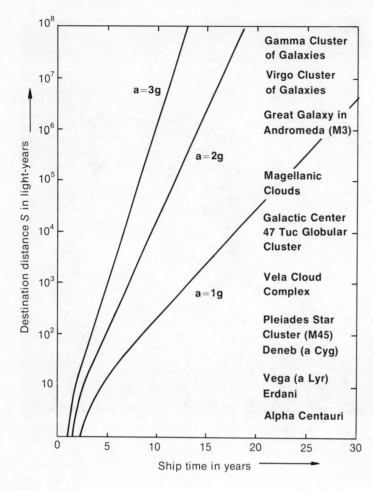

Figure 2–13 An illustration of the potentialities of time dilation in interstellar space flight. A space vehicle is imagined which has uniform acceleration of 1 g, 2 g, or 3 g to the midpoint of its voyage and a uniform deceleration thereafter. It is seen that immense distances—millions of light-years and more—could be reached by such vehicles during the lifetime of its crew. Yet the time passed on their home planet during the same voyage would amount to millions of years as measured by clocks there. (Reprinted with permission from Shklovskii, I. S., and C. Sagan, Intelligent Life in the Universe. San Francisco, Holden-Day, Inc., 1966. Courtesy of Pergamon Press, Ltd., London.)

man and a woman drawn to scale and standing before a schematic drawing of the spacecraft. A coded digital diagram indicates the position of our solar system with respect to certain pulsar star sources, and a drawing of our solar system indicates that the spacecraft originates from the third planet away from the sun (Fig. 2–15). All of this information could

© *by Walt Kelly*

Figure 2–14 © 1959 Walt Kelly. Courtesy Field Newspaper Syndicate.

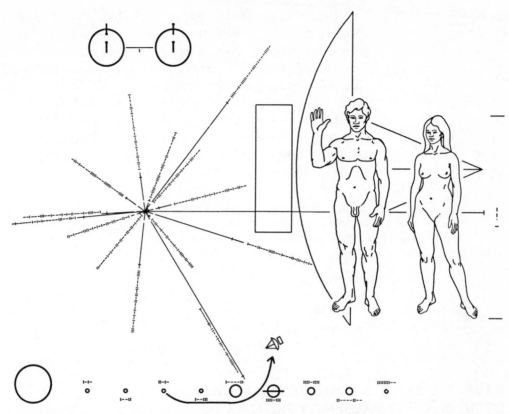

Figure 2–15 This message to unknown aliens starts with units of time and distance derived from minute changes in the energy levels of hydrogen atoms, upper left, the most common in the universe. In going from a state in which the spin of the electron and the spin of the proton are aligned, right atom, to one in which they are opposed, left atom, a hydrogen atom emits radiation at a wavelength of 21 centimeters every 1.4 billionth of a second. At extreme right, between the tote marks representing the height of the humans, appears the binary number 1 – – (which can also be written 100); this is 8 in the decimal system. Multiplication of that 8 by the 21 centimeters already established as a length unit gives a height of 168 centimeters, or about 5 feet, 6 inches. This number will also serve as a cross check for the aliens; if they have the message, they will have the spacecraft and can judge from that how big humans are. At left center is a polar projection of some object centered among 15 other objects. The binary numbers with each are too accurate to represent a distance; the only number that could be known with such accuracy would be the periods of pulsars. The solid line to each would then be its relative distance. Once the aliens identify the pulsars, they can pinpoint the origin of the space-craft. And because pulsars slow down with time, they can also determine when Pioneer was launched. Finally, to help locate us even more accurately, the solar system is sketched along the bottom of the plate. Here the binary 1's must obviously represent some unit other than 21 centimeters; in this instance, they indicate a tenth of the diameter of Mercury's orbit. The binary number near each planet gives the multiple of this unit for that planet. Pioneer, having left the third planet and swung by the fifth (Jupiter), has its dish antenna still aimed back at the earth. (From Sagan, C., et al.: A message from earth. Science *175*:881–884, Feb. 25, 1972. Copyright 1972 by the American Association for the Advancement of Science. Courtesy Carl and Linda Sagan and Frank Drake, Cornell University.)

be decoded by an advanced civilization to indicate the place of origin of the spacecraft. Furthermore, the instrumentation and electronics of the spacecraft itself will indicate to other beings the level of earth technology.

Technological civilizations elsewhere may contain beings

vastly different from ourselves in physical characteristics, and they are certainly not likely to speak a familiar language. In communicating with other beings by radio astronomy, scientists have developed a binary syntax (a mathematical language) that could be decoded by other intelligent beings. What is the purpose of attempting to communicate over such vast distances? Civilizations elsewhere may be far more advanced than ours; they may have long ago mastered such earthly problems as war, pollution, and overpopulation. Such civilizations may hold for us the very key to survival.

Who knows for certain? Who shall here declare it?
Whence was it born, whence came creation?
The gods are later than this world's formation;
Who then can know the origins of the world?

None knows whence creation arose;
And whether he has or has not made it;
He who surveys it from the lofty skies,
Only he knows—or perhaps he knows not.

The Rig Veda, X. 129

HAVE THE EARTH'S CONTINENTS ALWAYS BEEN IN THEIR PRESENT POSITION?

We shall not consider general aspects of geology or the origin of the earth,* but shall now turn to a contemporary problem in the earth sciences.

The nineteenth century view of the evolution of the earth can be summarized as follows: starting as a hot liquid molten body the earth radiated heat, cooled, and formed the continents, which became frozen in approximately their present positions. There were those who as early as the seventeenth century discussed the possibility that the continents were not always in their present position. In 1620 Francis Bacon suggested that from similarities in the shape of their coastlines, the western hemisphere was once joined to Europe and Africa. Some nineteenth century biologists felt that the early stages of evolution and speciation among plants and animals were difficult to explain by assuming the present arrangement of continents and would require the existence of land bridges across some of the oceans. Late in the nineteenth century the Austrian geologist Eduard Suess noted a close correspondence between geological formations within the southern hemisphere. He proposed that during the early history of the earth, the southern continents were joined together into a single continent which he termed Gondwanaland. In 1912, Alfred L. Wegener,

---

*For such a description see Gamow, 1963, or Asimov, 1966.

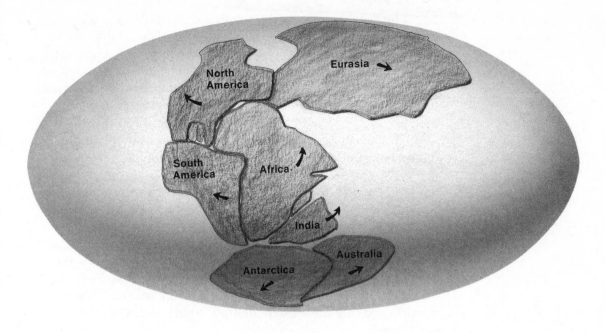

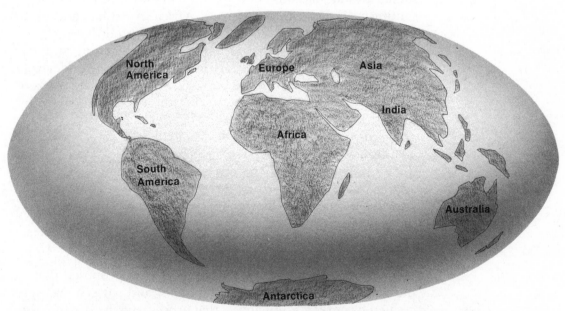

Figure 2–16   Single supercontinent presumed to have existed some 150 million years ago, would have re-sembled that depicted in the map at top. A present-day map appears at the bottom. In both maps the distortion of the continents is a result of the projection employed. (Adapted from Wilson, J. T.: Continental drift. Sci. Am., April, 1963, p. 97.)

finding similarities between fossils and rocks from opposite sides of the Atlantic, suggested that 200 million years ago, all the continents were massed together into one large land area, which he called Pangaea.

The idea that the earth's present continents were once joined into one or two great land masses and later "drifted" to their present position has existed for some time, but concrete evidence to support this theory did not exist until recent years. A book about earth science, first published in 1941 and revised in 1959, contains no discussion of continental drift, and as recently as 1963, J. T. Wilson spoke cautiously of this theory: ". . . I am not presenting an accepted or even a complete theory, but one man's view . . . ideas are rapidly changing and developing."

The historical development of theories of continental drift provides an excellent example of the long and patient accumulation of evidence that must occur before scientists accept a major revolutionary theory. Only recently, after years of debate and data gathering, did most scientists agree that the present continents were once joined into two great land masses—Gondwanaland in the southern hemisphere and Laurasia in the north—and that these later "drifted" apart to their present positions.

What evidence contributed to the development of this theory of continental drift and to its final acceptance? First, similarities in glaciation and fossil plants were found in the southern parts of South America, Africa, Australia, and in Antarctica. Studies of the earth's crust indicated that the present continental blocks, although relatively strong and approximately 100 km thick, are less dense than the underlying mantle, and thus, tend to be buoyant. After mapping areas of thermal activity, scientists found that the earth is encircled by distinct long and connecting belts of thermal activity. These are areas of high volcanic and seismic (earthquake) activity, heat flow, and other observed movements or displacements of the crust called faults. The directions of movements at these active zones were determined by the study of seismic waves from earthquakes, and evidence indicated that deep within the mantle, convection currents of heat were rising in some areas and sinking in others.

New evidence became available from oceanographic studies, and in the early 1960s marine geologists proposed that the great underwater oceanic mountain and rift systems were created by currents of material that arose from deep within the earth's crust and then spread outward over the ocean floor. In general, such rises separated areas of continents which were moving apart. Other zones existed, primarily in the ocean trenches, where sinking occurred and areas on both sides converged. Physicists studying magnetic properties of ancient rocks found that their magnetic orientation had been "frozen" into the rock in early times. They were then able to reconstruct the probable position of these rocks in relation to the magnetic pull of the earth during its early history. These studies also indicated that the continents had moved over the surface of the earth. Scientists were then able to map the general locations and time-span of such movements from magnetic study of both sides of submarine ridges.

In 1964, Edward Bullard and others at Cambridge University utilized a computer to find the optimum fit of continents within a single land mass. Finally, Patrick Hurley and others at the Massachusetts Institute of Technology found that the age of rocks on both sides of the Atlantic drift as measured by radioactive decay techniques were very similar. By 1966, most of the principal objections to the theory of continental drift had been overcome by the accumulating evidence.

It is believed that when Gondwanaland broke up, one part, in the form of the Indian subcontinent, moved northward, pushing up to and colliding with Eurasia. The thrust of India into Eurasia caused the buckling and pushing up of large land masses, forming the Himalaya Mountains.

The contemporary view of the origin of the continents is as follows: about 150 to 200 million years ago a single supercontinent called Pangaea existed with connections between Asia, Europe, North America, South America, Africa, Antarctica, India, and Australia. This single large land mass was like a plate floating on top of the plastic mantle of the early. In areas of rising thermal convection in the mantle, rifts began to appear. The supercontinent split first into two sections and then into others, which slowly drifted apart until reaching the present state. These continental plates continue to slowly move and their movements can be measured and mapped today (Fig. 2–16).

■ **REFERENCES**

Abell, G.: Exploration of the Universe. New York, Holt, Rinehart & Winston, 1966.
    (Introductory textbook format on astronomy, earth, planets, stars, etc.)
Allem, T.: The Quest: A Report on Extraterrestrial Life. Philadelphia, Chilton Book
    Co., 1965. (Popular nontechnical account of the search for life outside the earth.)
*Asimov, I.: The Universe: From Flat Earth to Quasar. New York, Walker & Co., 1966.
    (Excellent nontechnical review of earth, solar system, stars, galaxies, and theories
    of the origin of the universe.)
Barnett, L.: The Universe and Dr. Einstein. New York, The New American Library,
    Inc., 1960 (paperback). (Einstein's theories presented in an easy-to-understand,
    readable account.)
Berrill, N. J.: You and the Universe. New York, Fawcett World Library, 1959
    (paperback). (Stimulating thoughts in natural philosophy — man's place in a space-
    time relativity universe.)
Burbidge, G., and F. Hoyle: Anti-matter. Sci. Am. *198*:34–39, April 1958. (Story of
    the discovery and nature of antimatter.)
Cade, C. N.: Other Worlds Than Ours. New York, Taplinger Publishing Co., Inc.,
    1966. (Origin of life on earth, evolution of intelligence, stars, planets, and
    extraterrestrial life.)
†Clark, A. C.: Childhood's End. New York, Harcourt Brace Jovanovich, Inc., 1963.
†Clark, A. C.: 2001 — A Space Odyssey. New York, W. W. Norton and Co., Inc., 1968.

*Recommended further reading
†Fiction

*Coleman, J. A.: Relativity for the Layman. New York, The New American Library (Signet), 1969 (paperback). (Good nontechnical explanation including examples.)

Dirac, P. A. M.: The evolution of the physicist's picture of nature. Sci. Am. *205*: 45–53, May 1963. (Account of historical development of physical atomic theory and speculations on future theories.)

Gamow, G.: A Planet Called Earth. New York, The Viking Press, Inc., 1963. (Nontechnical introductory survey of the origin and evolution of the earth, its surface and atmosphere, and the moon.)

Gardner, M.: Can time go backward? Sci. Am. *216*(1):98–108, 1967. (Intriguing discussion of the symmetry of time and its implications for cosmological theories.)

Gatland, K. W., and D. D. Dempster: The Inhabited Universe. Greenwich, Connecticut, Fawcett Publications, 1963 (paperback). (Includes readable discussion of expanding universe, origin of life, and relativity, as well as the possibility of extraterrestrial life.)

Haber, H.: Stars, Men, and Atoms. New York, Washington Square Press, 1966 (paperback). (Cosmology, planetary evolution, the universe—finite or infinite?)

Hafele, J. C., and R. E. Keating: Around-the-world atomic clocks: observed relativistic time gains. Science *177*:168–170, 1972. (Technical paper—describes experimental evidence supporting Einstein's theory and relativity of time.)

Hannes, A.: Anti-matter and cosmology. Sci. Am. *216*(4):106–114, 1967. (Discussion of a new cosmology based on both matter and anti-matter.)

Hoyle, F.: The Nature of the Universe. New York, Harper & Row, 1960.

Is there life on Mars—or beyond? Time, December 13, 1971, pp. 50–58. (Concise, readable review of current ideas on extraterrestrial life and problems of communicating with intelligent beings outside our solar system.)

Ley, Willy: Watchers of the Skies. New York, The Viking Press, Inc., 1966. (An informal history of astronomy from Babylon to the space age.)

Message from mankind. Time, March 6, 1972, p. 68. (Short description of the Pioneer 10 spacecraft voyage beyond the solar system carrying a message to any intelligent life that may discover it.)

*Murchie, G.: Music of the Spheres. Vol. I, The Macrocosm: Planets, Stars, Galaxies, Cosmology. Vol. II, The Microcosm: Matter, Atoms, Waves, Radiation, Relativity. New York, Dover Publishing Co., 1967 (paperback). (Deals interestingly with most of the topics mentioned in this chapter.)

*Parsegian, V. L., et al.: Introduction to Natural Sciences. Part I: The Physical Sciences. New York, Academic Press, 1968. (An excellent introductory textbook including good illustrations.)

Sachs, R. G.: Time Reversal. Science *176*:587–597, 1972. (The notion that the direction of time flow is not knowable appears to be upset by recent experiments.)

Sagan, C. L., L. S. Sagan, and F. Drake: A message from earth. Science *175*:881–884, 1972. (More complete description of Pioneer 10 spacecraft.)

Sciama, D. W.: Modern Cosmology. New York, Cambridge University Press, 1971. (Review of current state of observational knowledge of universe.)

Sci. Am. *195*(3): September 1956. (This issue is devoted to articles concerning the universe, cosmology, etc.)

*Shklovskii, I. S., and C. Sagan: Intelligent Life in the Universe. Dell Publishing Co., 1968 (paperback). (A fairly comprehensive review including the general nature of the universe, stars, cosmology, origin of life, possibilities of extraterrestrial life, etc.)

Sullivan, W.: We Are Not Alone. New York, The New American Library (Signet), 1964 (paperback). (International nonfiction prizewinning account of why scientists believe there are intelligent beings on other planets and how we can contact them.)

Wahl, A. C.: Chemistry by computer. Sci. Am. *222*(4):54–70, 1970. (The use of computer in the construction of quantum mechanical models of atoms and their predictive ability.)

Wiley, J. P., Jr.: Intelligent life in the Universe. Natural History, February 1972, pp. 50–53.

Wilson, J. T.: Continental drift. Sci. Am. *208*:86–100, 1963.

---

*Recommended further reading

PROBLEMS

1. The theory of time-relativity was tested by flying an atomic clock around the world in an airplane. How does

this clock work and why did it support the theory of relativity?

2. If the universe was formed at one time from a "big bang," why are the furthest stars from us considered to be the oldest? Shouldn't everything be approximately the same age?

3. What is the effect of increasing velocity on the mass of a body?

4. Is the universe infinite? If not, where does it end?

5. Is there a specific time when the universe began?

6. Can theories about the nature and formation of the universe be proved? If not, how can we be sure they are true?

7. Do rocks or soil samples collected on the moon support any of the theories concerning the formation of the universe?

8. What does science hope to achieve by speculation about the origin of the universe?

9. If the universe began with the explosion of a superdense nucleus, why are the planets disposed in such an ordered solar system?

10. Do the findings of relativity theory invalidate the classical laws of physics? Why or why not?

11. Describe some theories for the origin of the universe not discussed in this chapter. This may involve a search of library references.

▰ BRAINSTORM

You have been appointed director of a team of scientists at a large government space research facility in Puerto Rico. The purpose of your team is to detect and communicate with other intelligent beings in the universe. How would you proceed? What techniques would you employ?

CONSIDER:
(1) Extraterrestrial beings may be extremely different from humans, both in form and in what they can perceive.

(2) It takes 4.3 years for light (traveling at 3×10^8 meters per second) to reach us from the nearest star, Alpha Centauri.

One evening a member of your team rushes into your office with a printed record of readings from the radiotelescope receiver that appear to fall into a regularly patterned series of pulses. A quick run of the record through your computer indicates that the pulses are a binary code representation of the number 3.1417 (the constant ratio of the circumference to the diameter of a circle). Should you direct your team to return a message? If so, what should the message be?

CONSIDER:
(1) Simply because this extraterrestrial life is intelligent, can we assume that it is friendly?
(2) We might be able to gain important lessons and knowledge from a civilization more advanced than ours.

3

THE ELECTRONIC AGE — CYBERNETICS, AUTOMATION, COMPUTERS, AND COMMUNICATION

Gigantic computer systems . . . computer networks . . . their inevitability cannot be accepted by individuals claiming autonomy, freedom, and dignity.

Joseph Weizenbaum

Automated cybernetic systems include devices that automatically perform sensory and motor tasks and replace or improve on human capacities for performing certain functions. They permit the mass production of cheap items and generally utilize methods of fabrication that make replacement more efficient or less trouble than repair. Increased automation will undoubtedly mean shorter working hours and fewer working days with consequently increased leisure time. Attitudes toward work, leisure, and social responsibilities will change greatly. Cybernated systems are having a profound impact on society, posing new problems and changing our way of life. The implications for society, as we shall see, are so far reaching as to challenge the very viability of the democratic system.

WHAT IS CYBERNETICS?

The word cybernetics derives from the Greek *kubernetes*, steersman or governor, thus indicating a control mechanism.

Cybernetics has been called the science of communication control in organized systems. Norbert Wiener, one of the early pioneers in this field, defines *cybernetics* as the study of common elements in the functioning of automatic machines and the human nervous system. It attempts to develop a theory covering the entire field of communications, machines, and living organisms. Obvious similarities exist between the functioning of the human brain and that of modern computers. Both involve memory, association, choice, and many other similar functions.

Cybernetics analyzes systems as a series of simple feedback control loops (Fig. 3–1). A desired condition of the system is selected by a *goal setting* process, entered into a *comparator*, and then tested against the *actual condition*, which is observed and reported by some process of *information feedback*. Any discrepancy or *disturbance* between the desired and observed conditions causes an *actuator* to modify the system and reduce the discrepancy.

The operation of a temperature regulating thermostat serves as a simple example. Here, the *goal setter* is the human who sets the desired room temperature on the thermostat. The *comparator* is a strip of metal (thermal contact) that receives the *information* (temperature) and expands or contracts, closing or opening an electrical switch (the *actuator*) which turns the *system* (furnace) on or off. A *disturbance* (change in room temperature) of the *actual condition* feeds back information to the *comparator* and the cycle is completed. The reader could construct similar analogies for the operation of many other information systems, from elevators to stock markets.

The basic concepts of cybernetics and feedback loops can even be applied to complex systems such as government. As E. S. Savas points out, urban governments display all the characteristics of cybernetic systems, including 1) feedback loops with a sophisticated and very complex governor or control device called government, headed in cities by the mayor; 2) information input loops to the mayor from his personal observations, public constituents, special interest groups, civil disorders, and

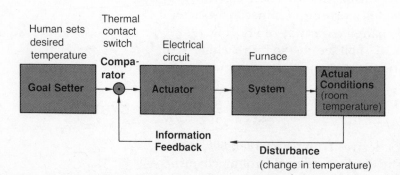

Figure 3–1 A basic feedback-control diagram. (Adapted from Savas, E.S.: Cybernetics in city hall. Science *168*:1066–1071, 1970.)

elections; 3) an actuator represented by the administration, which activates and guides the process by which broad goals are translated into specific objectives; 4) a goal setting system in the form of an executive branch; and 5) disturbances which can be classified as social, economic, or natural, which affect the system or city itself.

In addition to feedback loops, systems may have feed-forward loops, which are subject to anticipatory control. Feed-forward loops anticipate the effect of a disturbance on a system and take action to counteract the disturbance before the latter can affect the performance of the system. Savas and others believe that utilization of computers with feed-forward mechanisms may have real advantages in governmental operation. Problems and disturbances may be anticipated in advance and corrected (feed-forward) before reaching crisis levels. Planners believe that computers and communications networks (see later in this chapter) may make it possible for citizens to participate more directly in the decision-making process on an everyday basis.

THIS IS THE DAWNING OF THE AGE OF COMPUTERS

Man has utilized tools for several million years to expand his physical abilities. He developed mathematics as a tool to expand his thinking ability, and then developed tools to make computation easier and faster. Following the early use of sticks, pebbles, or knots on a string, man developed the abacus to solve mathematical problems. At the end of the nineteenth century, mechanical calculators gave birth to machines that could record information permanently on punched cards which could be read by other machines. These electrically driven mechanical calculators contained many movable parts and there were physical limits to their speed of operation. Between 1939 and 1944 Dr. Howard Aiken of Harvard University developed a machine called the *automatic sequence controlled calculator*, which embodied many of the principles of today's electronic computers.

In 1946 Drs. John Mauchly and J. P. Eckert of the University of Pennsylvania developed the first true electronic computer. This giant machine, named *ENIAC*, filled the entire basement of the engineering school at the University of Pennsylvania. It could find the answer to a problem in two hours that would have previously taken a hundred engineers a full year to solve. It used vacuum tubes to do its calculating work instead of mechanical gears or electromechanical switches, and it employed a form of machine logic that could solve complete problems by making decisions or choices as it went along. Early in 1960 Sperry Rand Corporation completed two models of a computing system much more powerful than any previously developed. This system, the *UNIVAC Lark*, could perform 28 billion calculations in two days and carry out the simultaneous solution of several problems at once. Amazingly, the development of this computer system occurred less than 15 years after the development of *ENIAC*, the world's first electronic computer. In the four seconds it would take a man to add two short rows of figures, the *UNIVAC Lark* system could perform one million additions or subtractions.

How Do Computers Work?

Computers may work with numbers that run into the billions, but they really respond to only two simple options—yes or no. Their mathematical operations are based upon the algebra of logic developed by the British mathematician George Boole more than a hundred years ago. Boole translated statements of logic into mathematical operations—logical propositions that could be proved to be either true or false. This was a relatively obscure branch of mathematics until 1938 when the American scientist Claude Shannon reasoned that the propositions of Boolean algebra are shown as either true or false very much as an electrical switch is either on or off. Modern electronic computers are really giant switching circuits. Each "decision" they make is essentially an on or off switch of an electric circuit.

Computers operate on the binary system of numbers, which recognizes only two digits, zero and one. Our everyday number system jumps one place when it reaches ten and another place when it reaches one hundred. The two-digit binary system, however, jumps every two digits. Thus, one is 1, two is 10, three is 11, four is 100, five is 101, and so on. In

BASE 10 NUMBER	BASE 2 (BINARY) NUMBER
0	0
1	1
2	10
3	11 (binary 2 + binary 1)
4	100
5	101 (binary 4 + binary 1)
6	110
7	111
8	1000
9	1001
10	1010
16	10000
32	100000
64	1000000

Figure 3–2 *A*, Binary numbers serve computers in logic, arithmetic, and coding functions. The array of binary numbers at left (shaded) shows that the system, which is based on 2, represents each new power of 2 by adding a 0. The same arrangement reappears at right in the binary version of the numbers 1 through 9: it shows, for example, that 111, representing 7, can be read from the left as "one 4, one 2 and one 1."

(Figure 3–2 continued on the opposite page.)

the simple language of the computer, if a switch is on, it stands for one, if it is off it stands for zero. Units of information transferred by information systems are called *bits* (a contraction of "binary digits"), which represent the uncertainty between "yes" and "no" or, in the binary number system, between 0 and 1 (Fig. 3–2).

The "switches" of most modern computers are really transistors — semiconductors that control the on or off flow of electrical current. In addition, some type of "*memory*" is present. One type, a magnetic core memory, is composed of thousands of ring-shaped cores threaded on thin wires. Electric current along one of the wires magnetizes a core; current passing in the opposite direction reverses the magnetic field. This is the same basic yes-no language of the computer, but the core can retain information as long as it is needed. All data are entered into the computer through the *input* on punched paper tape, punched cards, or magnetic tape, and the computer "reads" the information. A *control unit* decides what to do with the information. The program, a series of instructions stored in the computer's memory, tells the processing unit of the computer exactly what steps it must follow and how to make decisions when several alternative courses are present. *Data storage* is the computer's memory, where original information, intermediate results, records, and programmed instructions are stored. Each storage location has an "address" so the computer can locate the information it wants. The speed of a computer is largely determined by how fast it can get information out of storage and deliver it to another part of the computer. The *processing unit* is where the computer does its arithmetic operations of addition, subtraction, multiplication, and division. The *output* delivers processed information, usually in the same way it was put in — on punched cards, paper tape, magnetic tape, or video screen (Fig. 3–3).

		BASE 10	BASE 2
Addition		3	11
		+6	110
		9	10.01
Subtraction		15	1111
		− 10	1010
		5	101
Multiplication		$8 \times 5 = 40$	$1000 \times 101 = 101000$
Division		$54 \div 6 = 9$	$101100 \div 110 = 1001$

Figure 3–2 B, Binary arithmetic involves only the manipulation of 0 and 1 hence is the basis of the extremely rapid calculating done by computers. (Adapted from McCarthy, J.: Information. Sci. Am. *215*(3):65–73, 1966.)

Figure 3-3 *A-D.* A typical computer installation (IBM/370 Model 135) includes an input unit (*A*), where information is fed into the computer on punched cards or by teletype; a control unit (*B*), which controls operations; a magnetic tape data storage unit (*C*); and an output print-out for results (*D*). (Courtesy IBM.)

For a computer to operate, it must receive a set of instructions called a *program*. The human operator must design a set of operating instructions for the computer so that every problem can be solved in a yes-no manner. Information prepared on punched cards can be entered into a computer at the rate of more than 1000 cards per minute. A magnetic tape unit provides storage of intermediate results or long-term information, or by transfer to telecommunications it can communicate with other computers (see later in this chapter). A typewriter can be used for output of short messages, or the operator can use it to signal when computations should start. An output printer may print 1500 lines per minute—much faster than the human ability to read. In addition, computers may receive and display information on cathode ray tubes (television-type screens). Information is entered into the computer by drawing on the screen with a "light pen" and the computer registers the line as a series of points with certain space coordinates. Diagrams, circuits, geometric shapes, and other graphic data can be entered into the computer memory. In a similar manner, the computer can display graphic information such as graphs or diagrams from a stored program of mathematical information.

In about 1960 methods were developed to combine individual transistor components into miniature "integrated cir-

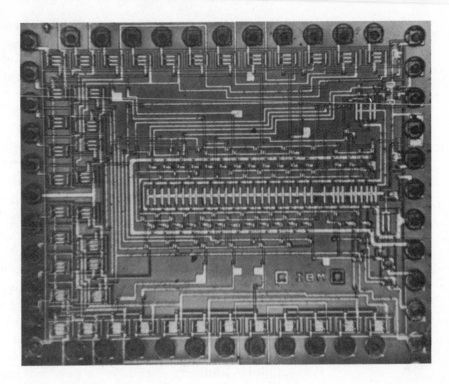

Figure 3–4 This single metal oxide chip, about 3mm square, contains circuitry which would have occupied a fairly large box of electronics in earlier years (a 32-channel multiplexor and a 5-stage counter). Such "large-scale integration" circuitry can be mass-produced in enormous quantitites by a process which in some ways resembles photography. This mass-production process will eventually reduce the cost of logic circuitry to a fraction of today's cost. (From Martin, J., and A.R.D. Norman: The Computerized Society © 1970. Reprinted by permission of Prentice-Hall, Inc., Englewood Cliffs, New Jersey.)

cuits." Today mass-produced integrated circuits contain more than 10,000 components on a small "chip" measuring less than 1 mm on each side (Fig. 3–4). These chips depend in part upon the new technology of metal oxide semiconductor (MOS) transistors.

The MOS transistor consists of two regions heavily "doped" with impurities called a source and a drain, and a gate electrode of metal separated from the underlying silicon by an intermediate insulator, usually silicon dioxide. When a negative potential is applied between a source and a drain, the conduction between them can be controlled by the modulation of a charge in the channel between them (Fig. 3–5).

Today virtually all small desk calculators and pocket calculators are designed with MOS circuits. Although MOS circuits are slower than the "bipolar" transistors used in large high-speed computers, MOS offers two advantages; lower power consumption and lower generation of heat. These advantages have enabled designers to increase the density of components, for example in the memory stores of computers. Furthermore, a new technology of circuit construction called SOS (silicon on sapphire), utilizing silicon layers only one micron thick deposited on substrates of synthetically produced sapphire, may overcome the inherent slowness of MOS circuits.

The use of ortho-ferrite crystals, smaller in diameter than a human hair, has resulted in a further miniaturization of computers. These thin crystals (2.5 mm square) can carry 10,000 bits of information; 15 million bits of information can be stored in about 16 to 32 cm³. With the small size of integrated circuits has come a decreasing cost per transistor, which may be less than one cent. Estimates indicate that by the early 1980s circuits with a density of one billion components will be produced at a cost of about .003 cent per component.

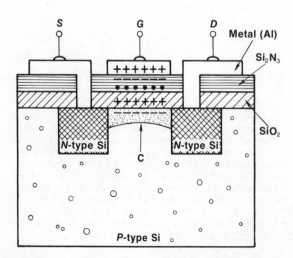

Figure 3–5 There are several kinds of metal oxide semiconductor (MOS) transistors, all of which function as rapid "on-off" electrical switches. In general, they contain an electrical conducting metal such as aluminum (Al), an electrical insulator such as silicon dioxide (SiO₂) and a semiconductor such as "N-type" and "P-type" silicon, which carry a current (flow of electrons) only after some minimum voltage has been applied.

In a typical type called an *enhancement type* transistor a positive charge (voltage) applied to the gate (G) induces a negative charge (electrons) to flow through a channel (C) between a source (S) and a drain (D). Up to a point, as the positive charge on the gate increases, the current flow through the channel increases. (From Diefenderfer, A. J.: Principles of Electronic Instrumentation. Philadelphia, W. B. Saunders Co., 1972.)

Computers operate at fantastic speeds, using transistors that turn off and on in as little as one billionth of a second. Nevertheless, more sophisticated use of computers, especially in space technology and weather prediction, often puts strains on the rapidity at which data can be processed. To achieve greater speed, IBM research labs recently developed an electronic switching device than can turn on and off in less than ten-trillionths of a second, more than a hundred times faster than transistors in most computers. Furthermore, it requires only about one ten-thousandth of the power of a transistor and produces much less heat.

IBM's development is based on a phenomenon discovered by the British physicist Brian Josephson of Cambridge University. Josephson determined that pairs of electrons would funnel through material that is normally an electrical insulator if it is thin enough and sandwiched between two superconductors.* If the flow of electrons through the insulator is low, there is little difference in voltage from one side of the insulator to the other, and, as Josephson predicted, if an external magnetic field is applied to the junction a voltage drop will appear. Computer scientists recognized that the presence or absence of voltage across a "Josephson junction" could be used to represent the same yes or no information conveyed by a transistor. There are many design problems to be overcome before this technique is applied to the manufacture of computers, but it will undoubtedly play a major role in the development of the high-speed supercooled electronic brains of the near future.

How Are Computers Used?

The uses of modern computers are almost limitless (see box insert). They include feeding livestock, helping make business decisions, mixing steel alloys, teaching students, planning cities, diagnosing illnesses, preparing payrolls, billing customers, tracking down criminals, designing automobiles, and guiding men to the moon. They now affect daily life in innumerable ways.

One prediction estimates that 75 per cent of all students graduating from college by the mid-1970s will need a working knowledge of computers in order to get a good job. Even today computer language courses are used to satisfy foreign language requirements at many universities. Furthermore, a survey of 250 computer experts from 22 countries predicted that by the year 2000 all major United States industries will be controlled by computers.

---

*Superconductance depends on the disappearance of electrical resistance in certain materials when they are cooled to near absolute zero, −273° C.

COMPUTERS—A FEW OF THEIR MANY USES

Basic Science	Study of subatomic and complex molecular structures, analysis of chromosome abnormalities, dynamic behavior of waves and fluids, ecological data and models, linguistic and semantic analysis of language, mathematical theory.
Technology	Industrial process and quality control, machining and manufacturing of tools from programs, production of computer components, structural design (e.g., bridges), manufacture of machine tools, flight simulation prior to construction of airplanes and rockets.
Transportation	Traffic control (trains, air, autos, mass transit), navigation (ships, airplanes, space vehicles), computerized map-making, ticketing, compute angles and grades for roads.
Business	Inventory sales and records, bank records, stock market records and analysis, marketing predictions, employment placement.
Government	Demography (population census), storage and analysis of crime records, military analysis of troop strength and strategy (war games), automated budgets, tax records, collecting weather data.
Education, Information, and Communications	Library storage and retrieval, "teaching machines" and transmittal, student registration and records, language translation.
Arts	Computer music (programs for Moog electronic synthesizer), computer graphic art, dramas written by plot manipulation, computer-programmed sculpture, prose and "poetry" composition.
Social and Entertainment	Date matching, gambling, child adoption placement, astrology and horoscopes, ticketing shows and sports events, computer chess.
Health	Medical record storage, symptom analysis, physical examinations.
Agriculture	Compute best combinations of crops and livestock.

Computers are now an essential tool for basic science. More than one-half billion calculations may be required for a single experiment in atomic physics. Quantitative data on ecological studies are usually analyzed by computer programs to compare populations of species and communities of organisms. In the social sciences computers store and compare information on various cultural traits or on population data. Computers provide a huge storage facility helping the scientist to file data for easy reference. They enable him to analyze information, compare facts and figures, find similarities and differences, classify information, and compare an otherwise large and unmanageable set of data. Robert Ledley of the National Biomedical Research Foundation in the United States has developed a technique whereby human chromosome abnormalities can be detected and analyzed by computer. Many human diseases or disorders have been found to be related to abnormalities in the nuclear chromosomes of living cells (see Chap. 5). Besides genetically inherited chromosome abnormalities, chromosome damage caused by certain substances or by ionizing radiation can sometimes be detected by photomicrographs. White blood cells are grown in tissue culture and then examined under the microscope for chromosome abnormalities. Ledley and others, instead of tediously taking thousands of photographs and examining them, have designed a computer program which can automatically scan a series of photomicrographs on a roll of film, feed them directly into the memory of the computer, and classify them by counting the total number of chromosomes, measuring their length, area, and other morphological features. This computerized method reduces the time required to examine 46 human chromosomes to 20 seconds—500 times faster than manual analysis by visual means.

This technique will be extremely useful for physicians who wish to examine the chromosomes of large numbers of people for a particular disorder. It should prove useful in screening new drugs and biological vaccines for possible chromosomal effects and in conducting large-scale studies of such matters as the effects of radiation and aging on chromosomes. Automatic visual analysis of micrographs, autoradiographs, and x-rays will be employed by biologists and research physicians for many other problems.

Space exploration would not be possible without modern high-speed computers. A man orbiting the earth at 18,000 miles per hour cannot wait weeks for calculations to be done by hand to locate his position at any moment. Instead, modern computer systems generate calculations in what scientists call "real time," in which calculations are continually updated and problems solved almost the instant they are stated. The continuous solutions to changing problems are printed out, with each solution reflecting the changes that preceded it. Computers are now required tools in tracking and computing the position of orbiting satellites and in the guidance and navigation of space vehicles. The black-and-white or color tones of television space-probe pictures are converted to mathematical notation and relayed to computers on the earth, which convert them back into processed pictures.

Computers are now being used to completely and automatically operate industrial facilities. In several petroleum refineries, computers read instruments, weigh hundreds of variables, and make the adjustments necessary to keep the industrial processes in balance. In other industries they coordinate materials, and inventory production scheduling more efficiently than ever before. They are used in some factories to automatically manufacture machine tools. Computers can work out detailed instructions and tell a machine how to produce a specified precision part. Then the machine, guided by the computer, turns out the part automatically (Fig. 3–6). In computer manufacturing itself, computers are used as tools to construct computer parts. They can be programmed to test by simulation every detail of an industrial or technical process before expensive construction of an actual plant is undertaken.

Computers rapidly handle large amounts of routine paperwork such as mail addressing, computation of paychecks, automatic processing of bank checks, inventory records, billing, and storage of credit information. At some department stores clerks feed the price of an article, along with the customer's credit plate number into a small countertop computer. At the store's central location, many miles away, credit is checked instantly, the bill recorded and subtracted from the account, and the item noted in the inventory of the store for reordering. Some bankers and businessmen have predicted the day when cash and checks will be completely outmoded. All purchases will be made by credit cards fed directly by computers to the bank, where charges will be made to the customer's account, and the bill paid.

In education, computers are already in widespread usage compiling report cards, registering students, and handling inventories of equipment. They are now increasingly being used in education as teaching machines. Harvard psychologist B. F. Skinner, an early pioneer in the development of teaching machines, has described their principles of operation and future potentials. The purpose of the teaching machine as stated by Skinner is "to teach rapidly, thoroughly, and expeditiously a large part of what we now teach slowly, incompetently and with wasted effort on the part of both student and teacher." Psychologists know that laboratory animals can be taught complex be-

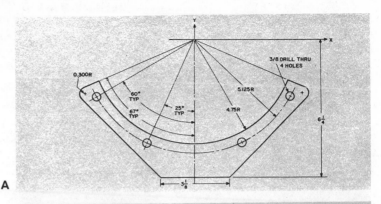

A

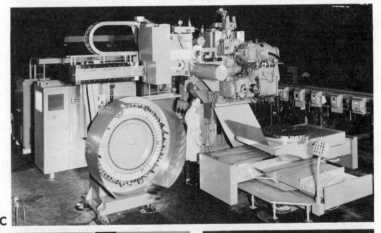

B

Figure 3-6 Machine tools can be controlled by computers. At the IIT Research Institute the process begins with a conventional engineering drawing of a part to be machined, in this case a small "radius plate" (*A*). From the drawing, the part programmer, writing in the APT (Automatic Programming for Tools) language, prepares a set of instructions that describe the part (*B*) and also the path to be followed by the tool. A computer calculates the detailed motions required to move the tool along that path, translating the programmer's word-symbols into numerical signals in the form of a punched tape. The tape controls the machine (*C*) in this case an "Omnimil" with 60 tools that are automatically interchangeable. A milling tool (*D*) shapes the part (*E*). (Courtesy IIT Research Institute, Chicago, Illinois.)

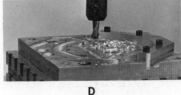

C

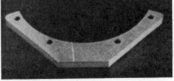

D   **E**

havior by the step-by-step reward or punishment method of operant conditioning. In a series of small steps, an animal can be led to the acquisition of complex forms of behavior; its behavior can be "shaped" (see Chap. 4).

Computer education technologies are being developed at many universities and private companies. A program devel-

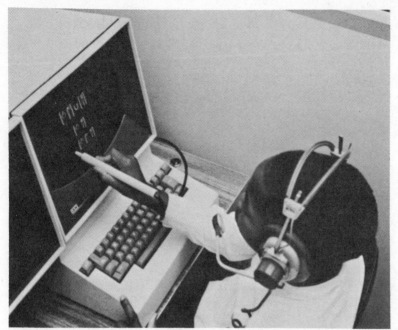

Figure 3-7 Children receiving mathematics lessons from computer terminals. The computer gives entirely individual instruction to many children at the same time, reacting differently to their different responses. (From Martin, J., and A.R.D. Norman: The Computerized Society © 1970. Reprinted by permission of Prentice-Hall, Inc., Englewood Cliffs, New Jersey.)

oped at Stanford University uses a TV display screen (for questions and information) and a typewriter keyboard (Fig. 3–7). The student types his name, and the computer determines the set of exercises the student is to receive on the basis of his previous performance. After a session with the computer, for example in arithmetic, the student is shown the number of problems correct, number wrong, the number incomplete, and his cumulative record. If the student's response is correct, his correct answer satisfies him and serves as its own "reward." The computer then moves on to the next question. If the student's response is incorrect, the computer will signify this and instruct the student to return to earlier information which he did not correctly learn. Students can work individually and have the advantage of knowing their level of competence without waiting for a test or final examination period.

Advocates of teaching machines argue that machines are in effect surprisingly like private tutors. There is a constant exchange between program and student, and like a good tutor, the machine insists that the material be thoroughly understood before the student moves on. Lectures, textbooks, and other devices, on the other hand, often proceed without making certain that the student understands the material. Slow students, free to learn at their own rate with the computer, may reach very high levels of competence. With computerized instruction students all achieve approximately the same level of competence in the end. Grades become more meaningful, reflecting accurately the amount of material the student has understood. Classes receiving computer instruction are generally far ahead when tested against classes not receiving computer instruction.

Within a few years, every student in the United States may have access to his or her own tutor in the form of a high-speed electronic computer. Even the architecture of schoolbuildings, Skinner points out, may have to be changed to make room for the computer. It may no longer even be necessary to continue a system of centralized schools. Mathematics and languages are the subjects currently taught best by computers, but the development of computers that can "understand" and speak English (or other language) (see later in this chapter) may pave the way for many more diverse areas of human knowledge to be taught by computers. Critics of teaching machines argue that this educational procedure is a cold and mechanical process that turns students into regimented, mindless robots.

For storage and dissemination of information, the computer may be the most important development since the printing press of Johann Gutenberg. The use of computers by libraries and information and retrieval systems is growing rapidly, from the first serious use of computers by scientific and technical libraries in about 1963 to current world-wide interest. A current project of UNESCO and the International Council of Scientific Unions (ICSU) includes plans for a worldwide system of science information. The Bell Telephone Laboratories have experimented widely with computers for information indexing, announcing, and retrieval purposes. Their computers are used to keep information up to date in a particular scientific field, and to keep people informed by conducting literature searches that are especially designed for a particular problem. For example, in one program at the Bell Laboratories, an author describes the interests of the readers to whom his report should be sent. The reader describes the type of reports he wants to receive. Computers match reports to readers and print address labels automatically. The Bell Laboratories' *Bell-Par System* provides publication, notification, and retrieval services for 40,000 technical papers selected from journals each year for announcements of specific new articles to individuals participating in the system, matching their computer-stored personal interests.

Another area of interest called "On Line" information systems serves the Bell Laboratories' three largest libraries and twenty additional library locations. Here a "real time library loan system" maintains an up-to-the-minute accountability for 120,000 books and 67,000 bound journals. It improves the total response of the libraries to the users by pooling the resources of geographically dispersed libraries, processing borrowers' reservations, automating clerical functions, and handling routinely more than 250,000 on-line library transactions per year. Within three or four seconds of a request, the computer generates the loan status of books or periodicals as on loan to a particular person; on order; missing; in transit between library locations; on the new-book shelf; or at the bindery. As Kenneth Lowry has said, "As an instrument for gaining information control and for keeping informed in given fields the computer has become almost an essential scientific tool."

Computers are even contributing to the area of art (Fig. 3–8). Victor Pickett, sculptor and associate professor of art at Old Dominion University in Norfolk, Virginia, has constructed an aluminum free-form sculpture from a computer program. Beginning with three-dimensional sketches, he worked out the mathematical formulas to translate a sculpture into a computer

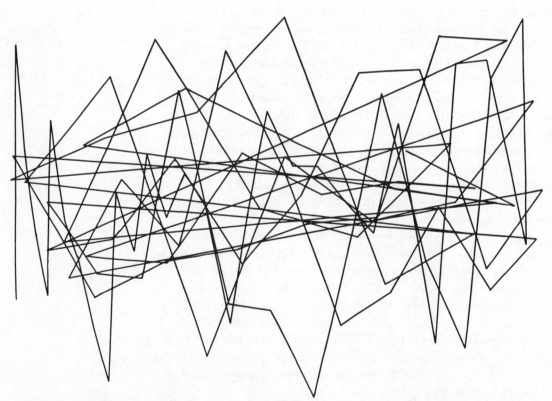

Figure 3–8 Computer-generated art includes works devised by A. Michael Noll of Bell Telephone Laboratories. In "Gaussian-Quadratic," the end points of each line have a Gaussian random distribution vertically; the horizontal positions increase quadratically. The pattern begins at left and is "reflected" back from right. (From McCarthy, J.: Information. Sci. Am. *215*(3):65–73, 1966. Copyright © 1966 by Scientific American, Inc. All rights reserved.)

program. The programmed information was transferred by punched-tape into the computer, and in the final stage the punched tape activated the controls of a machine tool which removed metal from an aluminum block, creating a three-dimensional sculpture.

Charles Dodge's recording of *The Earth's Magnetic Field* provides an intriguing example of the application of computers to the field of music. Electrically charged gases emanating from the sun (the solar wind) impinge upon the earth's magnetic field. Dodge recorded the levels or magnitude of the solar wind at specific time intervals with a sensitive instrument. The level of readings on the instrument was converted to musical notes. Almost 3000 readings for a one-year period were compressed into eight minutes of musical time by the computer.

In the area of criminal investigation, large amounts of data on criminals from all areas of the country can be stored in a central computer. A policeman stopping a traffic violator may request information on the person through his radiophone to headquarters and within seconds receive a report from the central storage computer in Washington, D.C., as to whether or not the person has a record of other violations.

In one California city, computers are helping to solve and prevent burglaries. Each week the police department gets an updated computer readout based on past burglary reports which predicts when, where, and how burglaries will most likely be committed that week. It allows police to review quickly all unsolved burglaries which fit into an apparent operational pattern of an arrested suspect. Patrol cars are assigned to areas most likely to be burglarized at a particular time. This has led to a 35 per cent reduction in the average number of burglaries per month. Future developments might allow computers to identify fingerprints and instantly check them against an up-to-date worldwide record.

In order to move information from where it is produced to where it is needed, computers need to communicate over long distances. Computers are now learning to use the telephone. Information can be transmitted over regular telephone lines from one computer to another at very high speeds. Giant computer communications networks hook up remote points thousands of miles apart. Thus, an airline reservations clerk, by pressing a button, can request a reservation on a certain flight for a certain day and in seconds a central computer thousands of miles away checks for reservations on that flight made by all the airline ticket offices around the country, establishes the availability of the space requested, and flashes back an answer. The reservations clerk then puts the information needed to make the reservation into the central computer, which updates its information instantly. This "real-time" system is adaptable to a large variety of business applications, including nationwide inventory and banking.

As computers become miniaturized and less expensive, they will be available to individuals, who will be able to use them for everyday tasks such as shopping and paying bills. Computer home hookups may eventually be as common as the telephone is now. People will want to learn to program, because knowing how to operate computers will be as essential as knowing how to drive a car.

Can Computers "Think"?

Analogies between advanced computers and the human brain are evident. Computers can set up goals, make plans, consider hypotheses, recognize analogies, and carry out various other intellectual activities. Norbert Wiener, one of the founders of the field of cybernetics, refutes the idea that machines cannot possess originality, and that nothing can come out of a machine that has not been put into it. He believes that machines can and do transcend the limitations of their designers and that "in doing so they may be both effective and dangerous." It is true that the machine may not be more "intelligent" than its human makers, but because of its fantastic speed of operation, human understanding of its mode of performance may not take place until long after its assigned task has been completed. In the game of checkers, computers show a marked superiority to players who have programmed them after only 10 or 20 hours of play. Programming of information may become so complex that the designer-operator will lose an effective understanding of the way in which the computer comes to its conclusions. The machine can operate at a pace so rapid that humans cannot keep up and may not know until too late when to turn it off.

Computers are now challenging man's ability to think in the game of chess. Claude Shannon first described how computers could be programmed to play chess in 1949. In addition to the purely intellectual challenge, scientists hope that the techniques developed in programming chess-playing computers can be applied to other types of problems that involve a selection of and search for alternate pathways, such as communications switching systems and electrical power grids. Furthermore, writing chess programs may provide insights into how the human brain works, how it analyzes patterns and quickly abstracts the important from the unimportant. Computers taught to play chess are now challenging each other in the game. The 1972 United States computer chess championship in Boston was won by a Control Data Corporation 6400.

Albert Zobrist and Frederic Carlson have described their program in which a chess-playing computer can be given advice by its human programmers. By accepting advice the computer can "learn" to generalize from particular mistakes. One of the most difficult tasks has been to translate specific chess knowledge into mathematical notations that computers can understand. Zobrist and Carlson's program is learning to correct its mistakes by taking advice from a chess tutor, Charles Kalme, a senior master in the International Chess Federation (Fig. 3–9). In the game of chess there are about 10^{43} possible board positions and about 10^{125} ways of moving pieces to reach these positions. Nevertheless, Zobrist and Carlson find that their programming of the computer has advanced to a point where they feel they can "talk" with the computer. The computer is no longer completely dependent upon expert programmers; it can acquire knowledge from chess experts rather than computer experts.

By dialing the proper telephone number, one can connect a teletypewriter to the computer, type in the proper password, and request execution of the chess program. The machine is

Figure 3–9 Man versus Machine game of chess is played between Charles I. Kalme, rated as a senior master, and the advice-taking chess machine, an IBM 370/155 programmed by the authors at the University of Southern California. Kalme, a member of the mathematics department at U.S.C., is the most highly ranked player engaged in the development of a computer-chess program except for Mikhail Botvinnik of the U.S.S.R. The U.S.C. program enables a chess expert with no previous computer experience to guide the machine to a more masterful game. Kalme's official rating is 2,445; machine currently plays at a sound novice level between 1,200 and 1,500. (From Zobrist, A. L., and F. R. Carlson, Jr.: An advice-taking chess computer. Sci. Am. 228(6):92–105, 1973. Copyright © 1973 by Scientific American, Inc. All rights reserved.)

then ready to play. After a move is typed in, the computer checks its legality, calculates a move in reply, and types out the move. The human opponent can request a printout of the chessboard and can give the computer advice if he wishes. The computer scans the board for the position of the pieces, entering between one and 3000 "snapshots" and storing them in its memory. It then "looks ahead," examining approximately 10,000 possible future moves and combinations in a period of fifteen seconds. Computers now play chess at a sound novice level, but they are learning rapidly and some experts believe that they will pose a challenge to international chess masters within the next five years. As Zobrist and Carlson point out, computers, because of their large memory capacity, could accept huge volumes of advice from teams of chess experts and in time know more than any one individual. They operate at high speeds, are not subject to lapses of memory or concentration, and will not be bothered by poor lighting or fatigue. "Once advice is given, the computer will not forget it and will never fail to apply it."

A research group at the Artificial Intelligence Laboratory of MIT headed by Marvin Minsky is currently studying the development of robot-devices with humanlike learning, vision, hearing, and manipulative capabilities. The robots will see via television, hear, understand English instructions, and have a sense of touch. They will be used to carry out tasks in deep ocean exploration, bomb demolition, and intelligence data collection.

The MIT group has programmed computers to make geometric analogies similar to those made on college entrance tests or intelligence tests in which a is to b as c is to ($d_1, d_2, d_3, d_4,$ or d_5). It is commonly believed that a computer can solve a problem only when every step of the solution is clearly specified by the programmer; but Minsky believes that it will eventually be possible for programs to apply the experience gained in solving one kind of problem to the solution of quite different problems.

Daniel Bobrow has demonstrated that computers can be programmed to understand a limited range of the English language. Such a computer will surpass most average people in ability to solve quantitative algebraic problems when stated in simple English. For example, his program called "Student" can rapidly solve the following problem: Mary is twice as old as Ann was when Mary was as old as Ann is now. If Mary is 24 years old, how old is Ann (Fig. 3–10)? Bobrow believes that given an enlarged English vocabulary memory, "Student" could understand most of the problems given in high school algebra textbooks. The remarkable thing is not that the computer can solve simple algebraic problems, but that it can understand human language.

(THE PROBLEM TO BE SOLVED IS)
(MARY IS TWICE AS OLD AS ANN WAS WHEN MARY WAS AS OLD
AS ANN IS NOW. IF MARY IS 24 YEARS OLD. HOW OLD IS ANN Q.)

(WITH MANDATORY SUBSTITUTIONS THE PROBLEM IS)
(MARY IS 2 TIMES AS OLD AS ANN WAS WHEN MARY WAS AS OLD
AS ANN IS NOW. IF MARY IS 24 YEARS OLD. WHAT IS ANN Q.)

(WITH WORDS TAGGED BY FUNCTION THE PROBLEM IS)
(MARY/PERSON IS 2 (TIMES/OP 1) AS OLD AS (ANN/PERSON)
WAS WHEN (MARY/PERSON) WAS AS OLD AS (ANN/PERSON)
IS NOW (PERIOD/DLM) IF (MARY/PERSON) IS 24 YEARS OLD.
(WHAT/QWORD) IS (ANN/PERSON) (QMARK/DLM)

(THE SIMPLE SENTENCES ARE)

((MARY/PERSON) S AGE IS 2 (TIMES/OP 1) (ANN/PERSON) S
AGE G02521 YEARS AGO (PERIOD/DLM))

(G02521 YEARS AGO (MARY/PERSON) S AGE IS (ANN/PERSON)S
AGE NOW (PERIOD/DLM)

((MARY/PERSON) S AGE IS 24 (PERIOD/DLM))

((WHAT/QWORD) IS (ANN/PERSON) S AGE (QMARK/DLM))

(THE EQUATIONS TO BE SOLVED ARE)

(EQUAL G02521 (ANN/PERSON) S AGE))

(EQUAL ((MARY/PERSON) S AGE) 24)

(EQUAL (PLUS ((MARY/PERSON) S AGE) MINUS (G02521))) ((ANN
/PERSON S AGE))

(EQUAL ((MARY/PERSON) S AGE) (TIMES 2 PLUS ((ANN/PERSON)
S AGE) (MINUS (G02521)))))

(ANN S AGE IS 18)

Figure 3-10 "Student," an English-reading program created by Daniel Bobrow, solves algebra problems. As shown here, Student restates a problem, analyzes the words in terms of its library of definitions and relations, sets up the proper equations, and gives the solution. The machine has invented the symbol G02521 to represent the X used in text. (Adapted from Minsky, M.: Artificial intelligence. *Sci. Am. 214*(1):19–27, 1966.)

Can computer programs "learn" through experience and improve themselves? The exploration of machine intelligence has just begun. As programs with a genuine capacity for self-improvement are developed, a rapid evolutionary process will begin and we may see the emergence in computers of what we call consciousness, intuition, and intelligence. Computers might then begin to surpass humans in their ability to think; human activity and aspiration, as Minsky says, would "be changed utterly by the presence on earth of intellectually superior beings."

In addition to the cyborgs discussed in Chapter 5, scientists have speculated on the future development of intelligent robots — mechanical men with a miniaturized supercomputer brain. In a science-fiction story Isaac Asimov describes future robots which are kept subservient to man only by the fact that they are all programmed to obey "The Three Laws of Robotics":

1) A robot may not injure a human being, or, through inaction, allow a human being to come to harm.
2) A robot must obey the orders given it by human beings except where such orders would conflict with the First Law.
3) A robot must protect its own existence as long as such protection does not conflict with the First or Second Law.

> Handbook of Robotics
> 56th Edition, 2058 A.D.
> (From Asimov, 1950.)

COMMUNICATION

Man surpasses all other animals in his ability to communicate. All organisms communicate with the environment; many communicate with each other; but only man communicates in terms of symbol systems. In addition to his complex patterns of speech he can now communicate knowledge from one generation to the next using written language. The technologies that have done more than anything else to revolutionize human communication in recent years are telecommunications. Telecommunications has been defined by a committee of the National Academy of Engineering as "any transmission, emission or reception of signals, written images and sounds or intelligence of any nature by wire, radio, visual or other electromagnetic systems including any intervening processing and storage."

The present generation of human beings has a capability of rapid and distant communication of information unparal-

TABLE 3-1   Milestones in Human Communication

Year	Event
Plus 1 million years ago to 200 years ago	Man communicates by runner, drums, smoke signals, etc.
1790s	Claude Chappé builds semaphore stations on top of hills throughout France in which mechanical arms are set by operator at coded relay points.
1844	Samuel F. B. Morse sends dot-dash message over single wire strung on poles between Baltimore, Maryland, and Washington, D.C.
1858	First transatlantic telegraph cable completed.
1876	Alexander Graham Bell demonstrates that the human voice can be electrically transmitted over wires.
1901	First radiotelegraph message sent across the Atlantic Ocean.
1923	Commercial transatlantic radio-telephoning introduced.
1945–1955	Rapid growth in commercial television.
1956	First high quality 36-channel trans-atlantic telephone cable installed.
1960	First operational laser.
1962	First active experimental communications satellite Telstar relays live television between US and Europe.
1970s	Networks of communication satellites each with a capacity of 5000 voice channels or 12 television channels being installed, video telephones, interactive television, and interactive computer terminals being developed.

leled in history. Early human communication between distant points was slow and of limited capacity. However, during the last two decades man's communicative abilities have accelerated so rapidly that we are now experiencing what might be termed a communications explosion based upon transmission of information via electromagnetic radiation (Table 3–1).

How Is Information Transmitted via Telecommunication?

Telecommunication information travels via electromagnetic radiation of differing wavelengths. Electromagnetic waves, in order to carry information, must be modulated (transformed) in various ways. The most widely used processes of modulation are amplitude modulation (AM), frequency modulation (FM), and pulse code modulation (PCM) (Fig. 3–11).

AMPLITUDE MODULATION

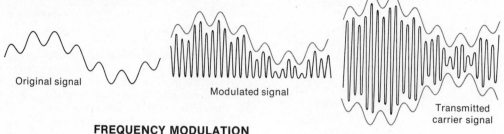

Original signal

Modulated signal

Transmitted
carrier signal

FREQUENCY MODULATION

Original signal

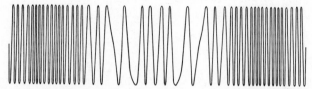

Transmitted
carrier signal

PULSE CODE MODULATION

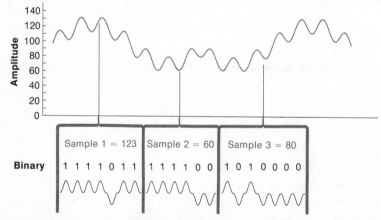

Figure 3–11 Communications channels generally carry information by electromagnetic radiation in three ways.

A, Amplitude modulation (AM) is commonly used for modulating commercial radio signals. The amplitude (height) of the original signal modulates (changes) the amplitude of a carrier wave of uniform frequency. In the radio receiver, the electromagnetic wave actuates a loudspeaker to produce sound waves.

B, In frequency modulation (FM), the amplitude of the original signal is converted into a wave of constant amplitude but differing frequency. As shown, when the amplitude of the original signal is large, the modulated signal is of greater frequency and vice-versa.

C, Pulse code modulation (PCM) has the advantage of carrying large quantities of information in a signal very resistant to external interferences (noise). The amplitude of the original signal is sampled 8000 times per second and coded into binary sequences of ones and zeros. At the receiver, the binary sequence of pulses is decoded to reproduce the original signal. Because this method uses a binary system, it can be used for transmitting computer information over long distances.

Channel	Channel Bandwidth (hertz)	Channel Capacity (bits per second)
Telephone Wire (Speech)	3,000	60,000
AM Radio	10,000	80,000
FM Radio	200,000	250,000
High-fidelity Phonograph or Tape	15,000	250,000
Commercial Television	6 million	90 million
Microwave Relay System (1200 Telephone Channels)	20 million	72 million
L-5 Coaxial-cable System (10,800 Telephone Channels)	57 million	648 million
Proposed Milimeter-Waveguide System (250,000 Telephone Channels)	70 billion	15 billion
Hypothetical Laser System	10 trillion	100 billion

Figure 3-12 Capacity of various communication channels can be measured according to Shannon's theory in bits per second, as indicated in the column at the extreme right in this chart. The numbers given for each entry are more or less rough estimates for a particular system (either currently in operation, proposed, or merely envisioned); in many cases it is difficult to ascribe even an approximate value for capacity in bits per second to an analogue channel because of the variability of parameters such as signal-to-noise ratio.

The *band width* of frequencies required to transmit depends on the type of information being communicated. Transmission of the human voice requires a channel frequency band from about 200 to 4000 hertz (cycles per second); that is, a band width of 3800 hertz. Voice information can be modulated to other frequencies as long as it carries the band width of 3800 hertz. Other channels of information require wider band widths. For example, American television stations are assigned band widths of 6 million hertz, and digital data produced by computers may require band widths of several million hertz (Fig. 3-12).

Four primary methods are currently used to carry telecommunication information over long distances. These are coaxial cable, microwave radio relay, wave guide, and satellite (Figs. 3-13 and 3-14). Computer data are sent over long distances by converting the wave form to digital data and transmitting them by pulse code modulation. The data are then reconverted to an analog wave. In this way they can be sent by microwave antenna or telephone line for thousands of miles.

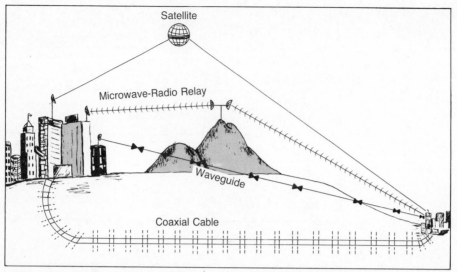

Figure 3–13 Four electrical techniques for transmitting a large volume of messages over a long distance are available at present. The newest technique involves the use of artificial earth satellites (*A*). The coaxial-cable system (*B*) still carries a large proportion of the communication traffic between cities in the U.S. The largest share of the intercity traffic in the U.S. is transmitted through the air by means of microwave-radio relay systems (*C*), with amplifying stations spaced some 20 to 30 miles apart. The waveguide technique (*D*), which has recently been perfected, will be able to carry more communication traffic than any other system currently available. Amplifiers (broken lines) are spaced 2 to 4 miles apart in the coaxial-cable system and 10 to 15 miles apart in the wave guide system. Microwave-radio relay horns are actually 10 to 15 feet in diameter. (Adapted from Miller, S.: Communication by laser. Sci. Am. *214*(1):19–27, 1966.)

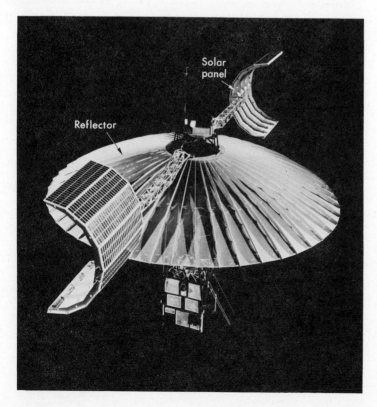

Figure 3–14 Applications Technology Satellite (ATS). This new spacecraft, the largest and most powerful communications satellite ever put into orbit, can receive signals from earth and transmit them to relatively inexpensive ground receivers. Total weight of the ATS is 1170 kg. Solar array consists of two semicylindrical panels supported by deployable booms. The reflector, which is nine meters in diameter, focuses radio frequency toward earth. (Photo courtesy of NASA.)

New Technologies in Communication

Communications channels of the future will probably utilize the new technology of lasers. In 1917 Albert Einstein proved that "stimulated" or controlled radiation could be obtained from an atom or molecule by stimulating it to a higher energy level and then controlling the released energy in a small electromagnetic field of the proper frequency. A working model of a laser (an acronym for "light amplification by stimulated emission of radiation") was first achieved in 1960. Information is carried by communication channels at differing frequencies of electromagnetic radiation. The capacity of a communications channel is proportional to the width of its band of frequencies. Because the laser utilizes electromagnetic waves in the visible region of the spectrum that includes a very wide band of frequencies, it is in principle capable of carrying many times the amount of information carried by lower frequency radio or microwave systems. The wide frequency band in the center of the visible region of the spectrum (where lasers operate) can theoretically carry a communications capacity 100,000 times greater than that of a typical microwave.

Given such communication potential, scientists in the Bell Telephone Laboratories and elsewhere are actively engaged in the development of laser communication systems. In an ordinary incandescent light bulb, light is emitted at many different frequencies, radiates in all directions, and is difficult to focus into a coherent beam which can be transmitted over long distances. In the laser, the number of possible frequencies emitted can be restricted and kept within a narrow spectral band. In a typical gas laser such as the helium-neon laser, a steady electric discharge is maintained through the gas mixture. This energy input excites the gas mixture, which emits light at specific wavelengths and frequencies (e.g., 632×10^{-9} m and 473 trillion hertz). A weak electromagnetic wave of the same frequency is directed through the tube and emerges as a more energized output wave with a plane wave front (as opposed to the spherical wave fronts of an incandescent light bulb). With a set of mirrors at each end of the tube, the light can be oscillated back and forth and amplified to a higher output level. Laser outputs up to millions of watts have been produced. The problem of transmitting communications over long distances by laser beams is now being investigated. As with other kinds of light, laser frequencies are attenuated by factors in the earth's atmosphere such as clouds and dust.

Newly developed channels for carrying light such as thin metal tubes, glass fibers, or ceramic liquid-filled fibers may provide the "pipes" to carry large volumes of information. Optical fibers for carrying laser light have theoretical band width capacities of millions of voice channels or thousands of television channels on a single beam. An optical light fiber telephoning system has already been developed and is being tested by the United States Naval Electronics Laboratory. Such telephones do not radiate external signals and are therefore "safe" channels of communication not subject to outside interference or monitoring. Optical fiber investigation now centers on the development of long fibers with low light attenuation, making fiber bundles into cables, and on interconnecting fibers. Our present combination communications networks—coaxial cable, microwave radio, wave guide, and satellite—can meet communications needs of the near future, and they have the advantage of diversification, but laser technology will undoubtedly play a large role in future communications networks.

The recent development of videophones, color videotape cassettes, video records and telecopiers will also have a large impact on communications technology and education. In one technology developed by the Phillips Company, a needle of laser light reads visual information from a spinning record. The record can store 45,000 individual picture frames, which can be played in regular or slow motion or even as still pictures. The advent of the telecopier has provided a way in which written, printed, or graphic (e.g., photos) material can be sent any distance over an ordinary telephone line. The telephone number is dialed, the receiver placed on a coupler, and within a few minutes an exact copy of the material emerges from the telecopier at the other end of the line.

Another newly developed communications technology is the electronic video file. Printed matter, photos, drawings, and other visual information can be stored on magnetic tape. For example, 750,000 pages of written material, enough to occupy 60 four-drawer file cabinets, can be stored on three small reels of magnetic tape. Any page of information among millions can be requested and displayed on a television screen within seconds to be read, or in a few more seconds, paper copies can be produced by the computer.

Miniaturized voice synthesizers such as *Votrax* have opened up new channels of communication between man and computer. Such voice synthesizers accept digital commands from a computer and convert them automatically into understandable language, enabling the computer to talk. They represent an analog of the human vocal system, receiving the counterpart of brain signals to the larynx, and duplicating human speech through the utterance of proper sounds.

Information Networks.   At present there are an estimated 7000 computers in the United States. They are used primarily for business processing, research, and education, and are distributed in a large number of separate centers and laboratories

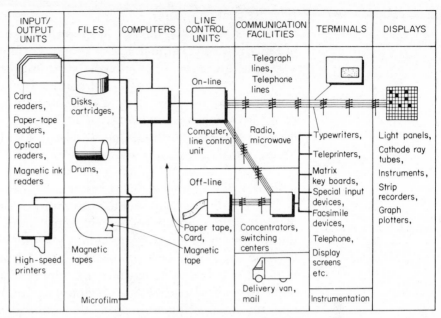

INPUT/OUTPUT UNITS	FILES	COMPUTERS	LINE CONTROL UNITS	COMMUNICATION FACILITIES	TERMINALS	DISPLAYS

Figure 3–15 The main types of devices that are attached to computers. (From Martin, J., and A.R.D. Norman: The Computerized Society © 1970. Reprinted by permission of Prentice-Hall, Inc., Englewood Cliffs, New Jersey.)

across the country; but they are being increasingly connected in complex *computer networks* (Fig. 3–15). For example, a medical researcher located in Honolulu can now search an index of the world's medical literature stored in a Maryland computer 5000 miles away. His request passes through a radio network in Hawaii, a telephone network across the Pacific via international satellite, and finally a time-sharing nationwide computer network, before reaching the medical information system in Bethesda, Maryland—all of this within less than five seconds. Ten seconds later, the requested set of literature citations begins printing out at the terminal in Honolulu. Meanwhile, other remote users of the medical information system receive simultaneous service. Use of this medical information system through a time-sharing network has been doubling every six months.

In 1972 the National Science Foundation (NSF) announced the initiation of an expanded research program to explore the resource-sharing potential of national computer networks in support of research and education. This program, called *EDUCOM*, sponsors collaboration between universities and other agencies in the use of computer technologies. Improvements in computer communications technology, data transmission, and switching procedures are leading to a rapid growth in the number and extent of computer networks. These include, for example, the *ARPANET* system developed by the Advanced Research Projects Agency of the Department of

Defense that connects dozens of independent computer centers from coast to coast. The NSF has sponsored the development of more than 30 regional networks among colleges and universities, including the *TUCC* in North Carolina, the *MERIT* in Michigan and the *UNI-COLL* in Delaware. Furthermore, many states are now developing statewide computer networks. The many advantages of "networking" include increased integration, resource sharing with reduced costs, greater availability to users, and thereby more efficient use.

Peter Goldmark describes the many as yet untapped potentials of the rapidly growing system of "interactive" cable television. These potentials include the ability of subscribers to order products, and participate in public opinion polls (with the responders' identity concealed . . . "if that is desirable"); provide warning of fire or burglary in the home; offer educational channels with an option for student response to questions; provide statistical data on viewer opinions; control the operation of heating, light, and other home devices according to the owner's program; read light, gas, and water meters, send the reading to the utility company computer, and return the bill to the user; and provide copies of printed material in the subscriber's home. Such systems could provide a pipeline into every home, putting everyone in the community in personal touch with everyone else including other homes, offices, and libraries.

In order to increase and speed up feedback and communication opportunities between government and citizens, a panel of the Committee on Telecommunications of the National Academy of Engineering has proposed the establishment of a pilot project to be set up in a typical city. A manned telephone system would operate around the clock and a computer would put citizens through a referral service.

Edwin Parker and Donald Dunn of Stanford University believe that the new technologies of cable television, especially two-way systems (video cassettes, computer information systems, and communications satellites) can now be developed into a nationwide "information utility," which would bring information and education to all citizens. Cable television systems enable subscribers to communicate data back to a computer at the source of the cable system. Such systems are called subscriber response systems (SRS). They permit a single computer to collect information from as many as 10,000 subscribers within two seconds. They can also record automatically whether the subscriber's television set is on and to what channel it is tuned. Parker and Dunn believe that cable television could provide a means to bring digital communications such as computer-aided instruction, information retrieval, and time-shared computer services into every home. Such systems could be interconnected into a national information network by micro-wave and communication satellite links.

The information utility they propose would be a communication network providing access to a large number of retrieval systems, information, entertainment, news, library archives, and educational programs to anyone at any time. It would provide the public with information about government opportunities, products, entertainment, knowledge, and educational services. Such a system might combine television, telephone, and typewriter, and function as a combined library, newspaper, mail-order catalogue, post office, classroom, and theater. Such an information utility network, they believe, could be available to most homes in the United States by 1985.

Regulations would have to be devised to guarantee freedom of speech and freedom of access to such a communications system from both the sending and receiving ends of the information utility. As the authors point out, information is power, and they believe that the variety of information could lead to an increasingly enlightened electorate with greater participation in the political process. On the other hand, we might ask, is not such a widespread complex network best controlled by a centralized authority such as government or large corporations? It will be a difficult task to guarantee that such centralized power over communications and information would not be abused.

Can Privacy Be Maintained in the Electronic Age?

Centralized computers with the capacity to store large amounts of personal data from widespread geographic areas represent an excellent means of assembling large amounts of information about an individual or group. Dossiers on every citizen could be collected through networks and stored in a centralized computer bank. Thus, there is growing concern that computers now (or will in the near future) constitute a dangerous threat to individual privacy. Furthermore, there is a growing potential for unauthorized use of computerized information. For example, individuals have gained unauthorized access to computerized bank or credit accounts of others and withdrawn large sums of money through the computer system. Can individual privacy and freedom be retained in the face of such computer potentials?

Interest is now growing in methods to protect computerized information from unauthorized access. Horst Feistel of the IBM corporation research group believes that privacy of computerized information can be guarded by ciphering or coding information that is fed to the computer. Only a few authorized persons with access to the special cipher or code could extract information from the computer. A properly coded information system would make it unlikely or impossible for an unauthorized person to withdraw or alter commands of the data in the computer.

Any type of information can be translated into a sequence of binary digits, enciphered with a binary number equivalent, and stored in a computer. By using certain techniques of random entry, a potentially undecipherable cipher can be entered into the computer. Such a system is already widely used by many governments, but deciphering of the information requires the same random sequence as enciphering and thus cannot be used more than once. Because of this restriction, the so-called *Vernam* system is usually reserved for top secret messages. More useful product-cipher systems, such as the *Lucifer* system (see Feistel, 1973) have now been developed and could provide a practical means of encoding computer data. Members of a centralized data bank community would retain privacy by a password authentication scheme. Members would carry their own private key (perhaps a sequence of binary digits recorded magnetically on a card), enter the key into a typewriter terminal, wait for a second, and begin typing their message to the computer. The message would be automatically enciphered in a cryptographic system, passed through an error-correcting code to reduce errors during transmission, and sent by telephone line to a receiving station. There, a password key would allow entry of the message into the computer center and the data would be entered into the storage bank of the computer. The primary weakness of Feistel's scheme is the problem of deciding who should be authorized to have access to the stored information. Could not magnetically recorded key cards be lost or stolen by unauthorized persons just as credit cards currently may be?

THE ELECTRONIC SOCIETY

The social impact of the three most widespread forms of telecommunications in the United States, the telephone, radio, and television, is already great. More than 90 per cent of United States homes contain at least one telephone, and an even greater percentage contain both a radio and television. An audience of 55 million people is watching 37 million television sets at any given moment of "prime time" in the United States.

How will our increasingly sophisticated technologies of information and communication challenge human freedom? George Gerbner states, "In a highly centralized mass production structure such as the kind characterizing modern communication . . . freedom is the right of the managers of the media to decide what the public will be told." The impact of mass communications on society results, as Gerbner points out, from the fact that the media (printing, television, and radio) provide the means of selecting, recording, viewing, and sharing man's notions of what is important or what is right, and tend to produce a new level of "common consciousness" or "modern mass publics"—heterogeneous social aggregates that never meet face to face, but share common messages from the media. Gerbner further states, "The real question is not whether the organs of mass communication are free, but rather: by whom, how, for what purposes, and with what consequences are the inevitable controls exercised?"

A study group of the US Office of Science and Technology (OST) in a recent report proposed that special FM radio receivers be placed in every American home to permit the government to communicate directly with citizens 24 hours a day. This proposal was part of a set of recommendations on how new technological developments could be used to meet the social needs of the country. The group proposed that the FM receivers be controlled by the government, even to the extent that citizens be unable to disconnect them. Recommendations of the report also included a "wired citizenry" and ultimately a "wired nation" system that would have information about police and court records and individual health records in a common computerized file system. The system as proposed would be operational by 1975 and would aid in alleviating problems of "urban unrest." The document acknowledged that there might be some concern about invasion of privacy. The proposal was rejected.

Can freedom of expression be maintained in the face of such advancing computer and communication capabilities? Much of being human depends upon the ability of self-expression and communication with other humans. Suppression of such communication can only contradict human dignity and integrity. Freedom of expression and a free flow of information promote man's search for truth and are essential to democratic decision making. They allow social change to occur after a confrontation of diverse ideas and help to avoid the need for violence and force.

Author Thomas Emerson points to the paradox that government interference in communications must be held to a minimum but that only government may be able to promote more even access to the increasingly centralized communications systems by minority groups and individuals. Regarding freedom of expression in the media, the US Supreme Court has stated that "it is the right of the viewers and listeners, not the right of the broadcasters which is paramount."

Joseph Weizenbaum of MIT argues that the computer has

had considerably less impact on society than is commonly believed. At least in their present state, systems such as automated airline reservations and computerized hospitals serve only a tiny affluent fraction of society, he argues, and do not have an impact on the general population. Gigantic computer systems and computer networks of information could be used effectively by centralized governments and by very large corporations for antihuman purposes: it is the responsibility of computer scientists to see that this does not happen. Emerson believes that man is still master of his technology and that it is the role of the conscientious scientist to determine to what uses the powerful computers will be put. Others have pointed out that if the scientist himself does not assume this responsibility, then society at large will be left to legislate against the possible evils of communication and computer technology.

The coming age of computer communication networks necessitates protection and regulation of use by both individuals and organizations, and has given rise to several important questions. Who should have priority access to networks? Who will be in charge of regulating and dispensing network time, and how will the privacy of network-stored information be guaranteed? Who will determine the individual's "right to know," or his access to information stored in central computer facilities? Information in computer memory banks may derive from individuals, corporations, organizations, or government. Who will have coded access to various types of information? Furthermore, how will the privacy of information concerning the individual himself be protected?

Computer intelligence and cybernetic technology are advancing so rapidly that within a decade or two it is doubtful whether most people will be able to understand the cybernated world in which they live. We may envision a small society of trained scholars capable of understanding and communicating with computers in a man-machine relationship while the rest of the populace, because of other interests or inclinations, will be involved in other tasks and will devote much of its time to expanded leisure activities. Donald Michael concludes that some type of control and restriction on freedom will be necessary to protect man from a contradictory world run by ever more intelligent and versatile computers: "The capabilities and potentialities of these devices are unlimited. They contain extraordinary implications for the emancipation and enslavement of mankind."

REFERENCES

†Asimov, Isaac: I, Robot. New York, Fawcett World Library, 1950 (paperback). (Collected short stories dealing with role of robots in future society.)
Busignies, H.: Communication channels. Sci. Am. 227(3):99–113, 1972.

† Fiction

Clark, J. O.: Computers at Work. London, Paul Hamlyn Ltd., 1970 (paperback). (Well illustrated descriptions of the many uses of computers; includes references for further reading in understanding computers.)

Coles, L. S.: Computers and society. Science *178*(4061):561, November 1972. (Advantages of computer networks are enormous, while evil side effects can be legislated against.)

Contemporary art by computer. Design *73*:18–19, 1971–72. (Sculpture from a computer-machine link.)

Dodge, C., et al.: Earth's magnetic field: realizations in computed electronic sound. New York, Nonesuch Records. (Computer music record.)

*Emerson, T. I.: Communication and freedom of expression. Sci. Am. *227*(3):163–172, 1972. (Modern communications technology creates problems for freedom of expression.)

Feistel, H.: Cryptography and computer privacy. Sci. Am. *228*(5):15–23, 1973. (The privacy of computerized information in central data banks can be protected by enciphering.)

Gerbner, G.: Communication and social environment. Sci. Am. *227*(3):152–160, 1972. (Mass communications has created a common consciousness, a mass public, and has important influences on society.)

Goldmark, P. C.: Communication and the community. Sci. Am. *227*(3):142–150, 1972. (Modern telecommunications can improve urban life and provide feedback between individuals, groups, and government.)

Green, S.: White house unit barred "Big-Brother" radios: government voice in all homes at all times. *International Herald Tribune*, November 2, 1972. (An Office of Science and Technology proposal to place government broadcasters in every American home is turned down.)

Greenberger, M., et al.: Computer and information networks. Science *182*:29–35, 1973. (National computer networks have many advantages and are growing rapidly.)

Hittinger, W. G.: Metal-oxide-semiconductor technology. Sci. Am. *229*(2):48–57, 1973. (Manufacture and utilization of MOS transistors in modern computers.)

How computers are changing your life. U.S. News and World Report, November 10, 1969, pp. 96–98. (The growing importance of computers in US society and economy.)

Inose, H.: Communication networks. Sci. Am. *227*(3):116–128, 1972. (Complex networks call for resourceful design.)

Ledley, R. S., and F. H. Ruddle: Chromosome analysis by computer. Sci. Am. *214*(4):40–46, 1966. (Chromosome abnormalities linked to human disorders can be detected automatically and rapidly.)

Lowry, W. K.: Use of computers in information systems. Science *175*(4024):841–846, 1972. (Bell Telephone Laboratories makes advances in information systems for library usage.)

McCarthy, J.: Information. Sci. Am. *215*(3):65–73, 1966. (Computers enable technology to adapt to human diversity.)

*McLuhan, M.: Understanding Media: The Extensions of Man. New York. The New American Library, 1964 (paperback). (Modern mass media are having profound impact on society.)

Michael, D.: Cybernation: the silent conquest. In M. Philipson, ed.: Automation: Implications for the Future. New York, Vintage Books, 1962. (Cybernation may radically alter the structure of society.)

Mighty New Servant of the Mind of Man. Sperry Rand Corporation, 1964 (pamphlet). (The operation, uses, and importance of computers.)

Miller, S.: Communication by laser. Sci. Am. *214*(1):19–27, 1966. (Principles of the laser and its development in communications.)

Minsky, M. L.: Artificial intelligence. Sci. Am. *215*(3):246–260, 1966. (Computers can be programmed to exhibit "intelligence" that may one day equal man's.)

Parker, E. B., and D. A. Dunn: Information technology: its social potential. Science *176*:1392–1399, 1972. (A proposed information utility could bring information to every home by 1985.)

Pierce, J. R.: The transmission of computer data. Sci. Am. *215*(3):145–156, 1966. (Large volumes of computer information are being sent over long distances.)

Pierce, J. R.: Communication. Sci. Am. *227*(3):30–41, 1972. (Advances in communications technologies have importance for society.)

Robinson, A. L.: Optical communications: specialized applications appear first. Science *182*:151–152, 1973. (Developments in optical fiber technology provide the first working light communication systems.)

Savas, E. S.: Cybernetics in city hall. Science *168*:1066–1071, 1970. (Cybernetics can be applied to improve urban government.)

Skinner, B. F.: Teaching machines. Sci. Am. *205*:91–102, November 1961. (Early experiments in teaching animals and humans by machine.)

---

*Recommended further reading.

Supercooled computers. Time, March 12, 1973, p. 48. (A short description of IBM's application of the Josephson effect to computer technology.)

Sutherland, I. E.: Computer inputs and outputs. Sci. Am. *215*(3):86–96, 1966.

Weizenbaum, J.: On the impact of the computer on society. Science *176*:609–614, 1972. (If scientists fulfill their responsibilities, computers will be used for humane purposes.)

Wiener, N.: Cybernetics. Sci. Am. *179*:14–19, November 1948. (An early description of the potential of a new area of science.)

*Wiener, N.: Some moral and technical consequences of automation. In M. Philipson, ed.: Automation: Implications for the Future. New York, Vintage Books, 1962. (Computers may surpass man in certain abilities.)

Zero Population Growth: A Teacher's Guide to Materials on Population. Denver, Colorado, 2 pg, PO Box 18291, 1971. (An excellent source of further material on population and resources. Includes bibliography of books, audio-visual materials, and population games.)

Zobrist, A. L., and F. R. Carlson, Jr.: An advice-taking chess computer. Sci. Am. *228*(6):92–105, 1973. (A computer "learns" chess from a master.)

*Recommended further reading
†Fiction

PROBLEMS

1. What are the size and cost of computers used in the space program?

2. Will men be replaced by computers? If so, what will the large number of unemployed persons do?

3. Will man be dominated by the technologies (e.g., computers) he produces?

4. With two-way interactive television for supplying education and services to society, will not many people stay at home, withdraw and lose much of their social interaction with others, to the point of not knowing the real from the imaginary? Will this not have a large effect on human relations?

5. Why do computers make mistakes?

6. Might it eventually be possible to synthesize "organic computers" or "biological units" of laboratory-cultured nerve cells with tremendous capacity?

7. What is a robot?

8. Is there any valid research going on in the field of extrasensory perception that could affect communications?

9. How does the "memory" of a computer work?

BRAINSTORM

Excerpts from the Universal Bureau of Investigation newsletter—for UBI classified employees only (2000 AD).

The UBI has completed installation of its "Every Home and Family Satellite Surveillance Network" (EHFSSN), which will facilitate the UBI's setting up of in-depth recording of family migrations, family overconsumption and black market activities.

The first prosecution brought under EHFSSN was filed in federal district court and involved a student in the Chy-Wals (Cheyenne-Walsenburg) megalopolis. The student was charged, on the basis of satellite observation by the EHFSSN, with trafficking in black market organic fruits and vegetables; receiving counterfeit Department of Transportation Computer Card Tickets (used for legal traveling, purchasing gas, auto repairs, etc.); and receiving illegal information from underground electronic networks, specifically literary and musical works prohibited as anticonformist.

(Adapted from ZPG, 1971)

4

THE BRAIN, ALTERED CONSCIOUSNESS, AND BEHAVIOR

By a careful cultural design, we control not the final behavior, but the inclination to behave—the motives, the desires, the wishes. The curious thing is that in that case the question of freedom never arises.

B. F. Skinner
Walden Two

In recent years, science, coupled with modern medical and electronic technology, has made impressive gains in understanding the basic structure and biological functioning of the human brain. Anatomists have traced the intricate networks of nerve cells that connect the sense organs and muscles to centers in the brain; cytologists have revealed the marvelous way in which electrical-chemical messages are transmitted along nerve cell pathways; physiologists have mapped centers deep within the brain that control emotions such as aggression, fear, and pleasure; behavioral psychologists have learned to control and "shape" human behavior; and scientists as a group (as well as the general population) have explored new levels of altered consciousness elicited by psychoactive drugs or meditation. Research will continue to add new knowledge to the field of psychobiology, but the question that now becomes increasingly important is how will man choose to use this new and powerful knowledge concerning the nature of his own mind?

THE STRUCTURE AND EVOLUTIONARY DEVELOPMENT OF THE HUMAN BRAIN

The human brain is the most complex structure in the known universe. This mass of ten billion nerve cells, weighing

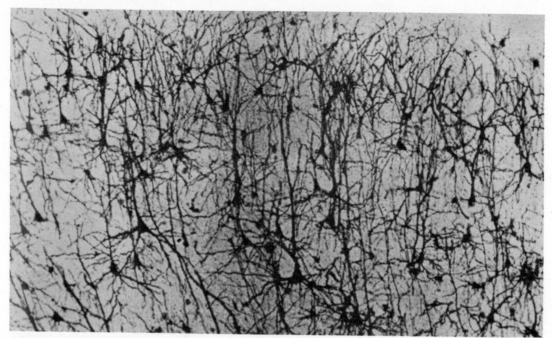

Figure 4–1 Cerebral cortex is densely packed with the bodies of nerve cells and the fibers called dendrites that branch from the cell body. This section through the sensory-motor cortex of a cat is enlarged some 150 diameters. Only about 1.5 per cent of the cells and dendrites actually present are stained and appear here. The nerve axons, the fibers that carry impulses away from the cell body, are not usually shown at all by this staining method. The photomicrograph was made by the late D. A. Scholl of University College London. (From Katz, B.: How cells communicate. Sci. Am. 205:209–220, September 1961. Copyright 1961 by Scientific American, Inc. All rights reserved.)

about 6.6 kg controls all actions and houses the center of what we call consciousness. There are about three times as many cells in one human brain as there are people on the earth. In some areas of the brain, one hundred million cells are packed into one cubic inch (Fig. 4–1). The outer folded and convoluted portion, the cerebral cortex (or cerebrum), occupies 80 per cent of the brain's volume. If the many infoldings of the cerebral cortex were flattened out, the tissue would cover an area 60 by 90 centimeters.

The cerebrum is the location of important sensory and motor areas (Fig. 4–2D). Sensations of touch and temperature, and impulses from muscles concerning general equilibrium reach the sensory area of the cerebral cortex. The motor area of the cortex sends signals for muscle contractions to various areas of the body such as the legs, trunk, arms, and face. Separate lobes of the cerebrum serve as centers for speech, hearing, and sight. The cerebrum, furthermore, is the center for language, abstract thought, and consciousness.

Behind and beneath the cerebral cortex is the cerebellum, which monitors bodily movements and smooths and coordinates impulses leading to muscular movements. It is the primary organ of motor coordination. The thalamus, located near the top of the brain stem, acts as a major relay station or switch-

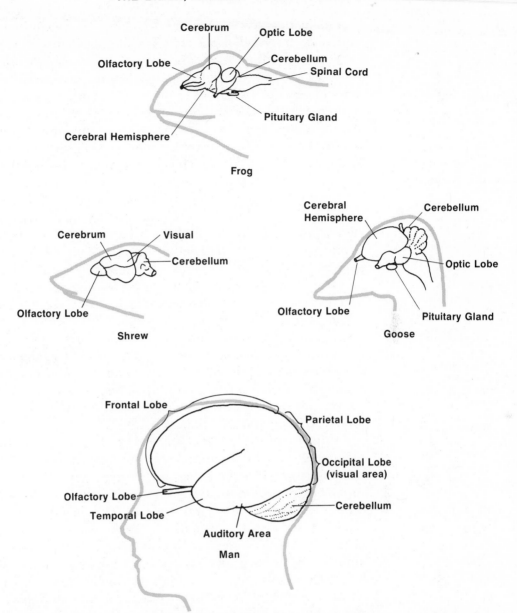

Figure 4–2 Location of important areas of cerebrum in frog, goose, shrew, man. (From Asimov, I.: The Human Brain: Its Capacities and functions. New York, New American Library, 1963.)

board connecting the complex web of nerve circuits. Sensory information is analyzed here and relayed to various areas of the cortex. In front of the brain stem, the hypothalamus constantly monitors vital signs such as body temperature, blood pressure, and heartbeat. Also, it houses the control centers of hunger, thirst, and sex, as well as such emotional reactions as fear and aggression.

An area called the reticular activating system (RAS), located between the top of the spinal cord and the thalamus and hypothalamus, acts as a major screening center which controls information relayed to the conscious mind. All incoming and outgoing communications appear to be channeled through the reticular activating system. Digestive contractions of the stomach, for example, may pass impulses to the brain that are screened by the RAS and do not reach levels of consciousness. In effect, then, the reticular activating system acts as a switchboard, deciding which incoming messages should be amplified and responded to, and which are to be minimized and ignored. It is important in controlling sleep and wakefulness.

The human brain is a product of a long evolutionary history. In tracing this history from lower animals to man, the most significant change is seen in the great increase in the size of the cerebral cortex or "higher center" (Fig. 4–2). In fish, there is essentially no cerebral cortex present; in amphibians (such as the frog) and reptiles, a small primitive cortex begins to emerge; and in birds we find the first nonolfactory (non-smell) functional cortex.

There have been differences in evolutionary emphasis among various vertebrates. In fish and amphibians smell is the primary sense and the olfactory lobes are well developed, but in most birds smell is relatively unimportant while sight is very important for flight behavior and food location; thus, the optic lobes are large and well developed.

Only in the evolution of one group of vertebrates, the mammals, does the cerebral cortex become a distinct structure with important functions. In some mammal types (such as rats) the cortex remained relatively small and undeveloped, but other groups, notably the primates, developed a greatly enlarged cerebral cortex. Primitive primates such as tree shrews originated about 70 million years ago. We may theorize that among these early primate types (e.g., shrews and tarsiers) larger cerebrums provided a higher "intelligence" or variety of learned behavioral responses to environmental stimuli. Consequently, in competition for food and other resources, those individuals with better developed cerebrums had a greater chance of survival and reproduction than those with smaller cerebrums. Following the emergence of primitive primates an evolutionary period of at least 60 million years led to a variety of larger-brained primates.

In the evolution of the higher apes and man we find the development of types with increasing cranial volume to contain the much enlarged cerebrum. *Australopithecus*, an apelike "man" present in Africa 2.5 to 3 million years ago, had a cranial volume of less than 500 cubic centimeters. (A newborn modern baby by comparison has a brain volume of about 300 to 400 cubic centimeters.) *Homo erectus*, roaming central Africa about 1 million years ago, had a cranial volume of nearly 800 cubic cen-

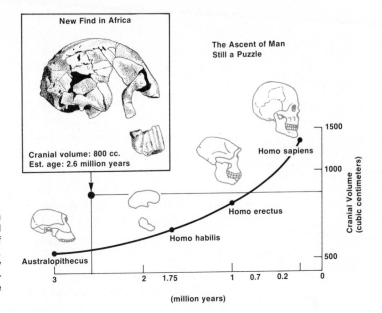

Figure 4–3 The gradual evolution of man's brain has been represented as a progression from the skull of Australopithecus to Homo sapiens. This progression is challenged by the recent discovery by Richard Leakey of a prehistoric skull near Lake Rudolf in East Africa.

timeters, and in an average modern man the skull surrounds a cavity of 1500 cubic centimeters. Richard Leakey's recent discovery of an African fossil skull 2.6 million years old with a cranial volume of 800 cubic centimeters indicates that relatively large-brained primates may have existed at a much earlier date than previously thought (Fig. 4–3).

As the cerebral hemispheres enlarged from the simple forebrain of the frog to the gigantic cerebral cortex of man, they also developed new connections with the brain stem and formed more intricate connections with the lower centers of the hypothalamus, the thalamus, and the reticular activating system. Thus, the evolution of intelligence may in part be due not only to the enlargement of the cerebral hemispheres but also to their better communication and coordination with the other "lower" centers of the brain. The greatest enlargement has occurred in the temporal lobes of the cerebrum, which participate in controlling speech, and in the frontal lobes, which control abstract thought. One important difference between man and the other primates is his greater capacity to plan and reason out the consequences of future actions.

PROBING THE BRAIN

Walter Hess of the University of Zurich received the Nobel prize in physiology and medicine in 1949 for his demonstration 20 years earlier that electrical stimulation of certain areas in the hypothalamus of cats could produce fear, anger, and reactions connected with bodily functions. His discovery opened up a new and highly fruitful area of research involving electrical stimulation and mapping of brain centers. Small metal electrodes measuring only one-millionth of an inch in diameter are inserted through the skull of an animal or man into a single nerve cell within the brain. A small electric current passed through these electrodes can stimulate the desired area of the brain to produce a reaction normally controlled by that particular area. For example, the limbic system, a center for the control of emotions, has been "mapped" in this manner. By stimulating one area of the limbic system, monkeys can be made excessively aggressive, or by stimulating a different area, they can be made extremely passive. Rats can be made to eat until they are three times their normal weight. Dr. Jose Delgado of Yale University found that stimulation in one area can activate an animal's appetite, and stimulation of another area can make the animal so uninterested in food that it would starve to death if continuously stimulated.

Electrical stimulation of the brain (ESB) has shown that brain centers exist for both pleasure and pain. Rats with electrodes implanted in the "pleasure center" of their brain will press a lever that delivers a self-stimulating electrical impulse to this pleasure center as often as 10,000 times an hour or until

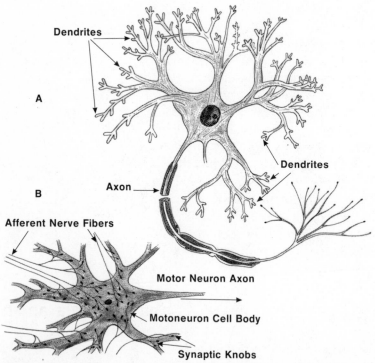

Figure 4–4 *A*, Motoneuron cell body and branches called dendrites are covered with synaptic knobs, which represent the terminals of axons, or impulse-carrying fibers, from other nerve cells. The axon of each motoneuron terminates in turn at a muscle fiber. (Adapted from Asimov, I.: The Human Brain: Its Capacities and Functions. New York, New American Library, 1963.) *B*, Network of multibranched neurons.

they drop from exhaustion (Fig. 4–4). Dr. D. O. Hebb of McGill University found that rats with electrodes placed in their pleasure centers, even when first starved for 24 hours or more and then placed in a cage containing food, would ignore the food, go to the lever, and begin the lever-pressing self-stimulation despite their hunger — in other words, they preferred the pleasure center to food. Furthermore, it has been found that different types of pleasure are represented by different regions of the pleasure center. One area serves as a hunger-reward center, and when stimulated will have the same satisfying effect on hunger as does food. Rats will actually bar-press and self-stimulate this center until they die of starvation. Other regions of the pleasure center appear to be centers for sex. Stimulation of this sexual pleasure center in a monkey can cause it to stay awake enjoying the experience for up to 48 hours, apparently preferring the experience to sleep. As Dean Wooldridge has said, "Satisfaction of the basic drives of hunger and sex appears to be simply a matter of the presence of electric current in the proper neuron circuits of the brain!"

Deep electrical stimulation of the brain has also resulted in the discovery of centers of rage, fright, and pain. Stimulation of such a "punishment center" in the hypothalamus results in the secretion of hydrochloric acid in the stomach and intestines and eventually produces peptic ulcers. By stimulating areas deep within the brain that control aggression, fierce bulls have been stopped in midcharge and cats have been shocked into dropping rats they were about to kill.

Most research on the pleasure and pain centers has been performed on laboratory animals, but research on electrical stimulation of the human brain is progressing. Stimulation of several centers in the human brain produces sensations described by patients as peace, relaxation, joy, or great satisfaction, while stimulation of other areas produces anxiety, restlessness, depression, fright, and horror. Electrical stimulation of certain areas in the brain of epileptic patients has induced memories in which the patient experiences a flight of images and vividly recalls past experiences in his life. With arrangements for self-stimulation of the pleasure centers, patients have been known to stimulate themselves repeatedly until reaching convulsions. Afterwards, however, they are relaxed, smiling, and happy. Properly placed electrodes in the human brain can sometimes bring dramatic but temporary relief from pain. To those who question the ethics of such human experimentation, proponents answer that such delicate operations are usually performed only on patients suffering from intractable pain, severe depression, or a psychosis that has not responded to any other treatment.

Since the demonstration of a pleasure center in the human

brain by electrical stimulation, scientists have speculated on the future possibility of human "elads" (electrical addicts) habitually addicted, not to drugs, but to the electrical stimulation of their pleasure centers by implanted electrodes.

In the past, society functioned to enhance individual well-being through promotion of the general welfare. However, elads equipped with modern technology could individually achieve electrical happiness without contributing to the general welfare or happiness of others in society. Will society permit such individuals to "drop out" and exist as a dependent subculture? Or does society have higher responsibilities that would justify its challenging the freedom and "inalienable right" of individuals to "liberty and the pursuit of happiness"? Will dictatorial governments legalize, take control of, and operate ESB centers where addicts receive regular ESB as long as they serve the wishes of the government? These are only some of the important questions that may challenge society as a result of our new knowledge of the human brain.

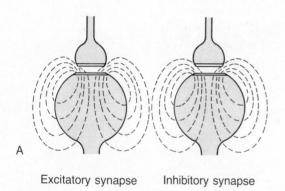

Excitatory synapse   Inhibitory synapse

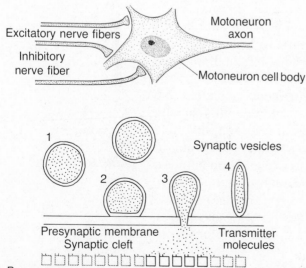

Excitatory nerve fibers

Inhibitory
nerve fiber

Motoneuron
axon

Motoneuron cell body

Synaptic vesicles

Presynaptic membrane   Transmitter
Synaptic cleft   molecules

Subsynaptic membrane

Figure 4–5 *A,* Current flows induced by excitatory and inhibitory synapses are shown at left and right, respectively. When the nerve cell is at rest, the interior of the cell membrane is uniformly negative with respect to the exterior. The excitatory synapse releases a chemical substance that depolarizes the cell membrane below the synaptic cleft, thus letting current flow into the cell at that point. At an inhibitory synapse current is reversed.

B, Synaptic vesicles containing a chemical transmitter are distributed throughout the synaptic knob. They are arranged here in a probable sequence, showing how they move up to the synaptic cleft, discharge their contents, and return to the interior for recharging.

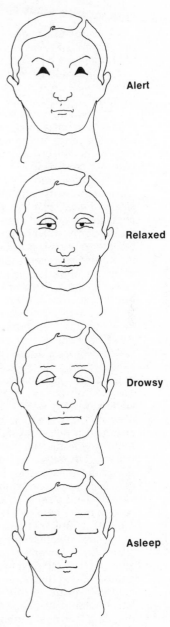

Alert

Relaxed

Drowsy

Asleep

Figure 4-6   Electroencephalograph (EEG) recordings indicate that the electrical activity of the human brain is typically rapid (waves of high frequency) during the alert state of consciousness and slows down progressively until a state of sleep is reached.

Figure 4-7   Self-stimulation circuit is diagrammed here. When the rat presses on treadle it triggers an electric stimulus to its brain and simultaneously records action via wire at left.

How are Messages Transmitted in the Nervous System and Brain?   The billions of neurons (nerve cells) that make up the human brain are composed of three main parts—the *cell body* containing the nucleus, branching *dendrites* which form antenna-like projections, and the long *axon* which extends from the cell body (Fig. 4–5*A*).

In a resting nerve cell, sodium and chloride ions are in higher concentrations outside than inside the cell membrane, while the concentration of potassium ions inside the cell is higher than the concentration outside. These differences in ionic concentration result in a difference in electrochemical potential (or voltage) of charged ions, where the inside of the cell membrane is negatively charged with respect to the outside of the cell by about 70 millivolts. However, when a nerve impulse is triggered in some way, the membrane becomes locally permeable and sodium ions pour into the cell, making the interior locally positive. This reversal of polarity of the membrane represents the nerve impulse which travels as a wave the length of the nerve cell axon while, in its wake, the permeability of the membrane decreases, restoring the normal polarity within a millisecond or less.

The multi-branched neurons are connected in complex networks (Fig. 4–5*B*). Junctions where the dendrites of one nerve cell intercept the cell body or axon of other neurons are called *synapses.* At the synapse a small space between nerve cells (synaptic cleft) acts as a barrier to the transmission of impulses from one cell to another. Electrochemical impulses coming from connecting neurons may be either inhibitory or stimulatory (Fig. 4–6*A*). But when incoming stimulations exceed inhibitions, a recipient neuron "fires" and an electrochemical impulse travels rapidly down the neuron's membrane momentarily altering the normal electrostatic equilibrium between the outside and inside surface of the neuron membrane.

Within enlarged terminal *synaptic knobs* are many tiny sacs or vesicles containing transmitter chemicals. When an electrical impulse reaches these vesicles it causes them to release their chemicals (Fig. 4–6*B*). According to current theories, inhibitory transmitter chemicals open the membrane to the flow of potassium but not to sodium ions; excitatory transmitter substances permit the free inflow of sodium ions, allowing the electrical impulse to cross the synapse to the next neuron. These transmitter chemicals involved in nerve transmission may include acetylcholine, norepinephrine, dopamine, and serotonin, and each may be active in different pathways of the brain. The neuron then returns to its normal state, the whole cycle taking only about one-thousandth of a second.

With the electroencephalogram (EEG), small electrode wires are taped to the outside of the subject's scalp and the sensitive instrument monitors the brain's activity in characteristic wave patterns (Fig. 4–7).

What Is the Chemistry of the Brain?

Man has known for thousands of years that very small amounts of certain chemical compounds could drastically alter the consciousness of his mind. The effects of one type of mind-altering drug were discovered by Albert Hofmann, a Swiss organic chemist, in 1943. Hofmann, experimenting with a chemical compound derived from a rust-fungus that grows on wheat, added some new chemicals to it and somehow absorbed some of the mixture into his system, perhaps through the pores of his skin. Shortly afterward he went home not feeling very well. He reported:

> On arriving home I lay down in a dazed condition with my eyes closed. There surged upon me an uninterrupted stream of fantastic images of extraordinary plasticity and vividness and accompanied by an intense, kaleidoscopelike play of colors. I felt a marked desire to laugh. I had great difficulty in speaking coherently, my field of vision swayed before me, and objects appeared distorted like images in curved mirrors. The faces of those around me appeared as grotesque, colored masks. I had a clear recognition of my condition, in which state I sometimes observed, in the manner of an independent, neutral observer, that I shouted half insanely or babbled incoherent words. Occasionally I felt as if I were out of my body. All acoustic perceptions (e.g., the noise of a passing car) were transformed into optical effects, every sound evoking a corresponding colored hallucination constantly changing in shape and color. . . .

Hofmann was aware that ancient tribes had used hallucinogens derived from mushrooms and cacti for thousands of years, but what surprised Hofmann was the extremely small amount of the chemical required for his "trip"—the dose was only about 100 micrograms or roughly one three-millionth of an ounce—the chemical was lysergic acid diethylamide (LSD). If such a small amount of a chemical could so dramatically influence the mind, causing hallucinations somewhat like those experienced by psychotics (the hallucinogens have also been described as psychotomimetic), perhaps mental instability and insanity were actually the result of tiny chemical imbalances in the brain.

Scientists hoped that by understanding the action of compounds such as LSD, the physical basis of mental illness could be found and certain types of mental illness could be counteracted or cured. Many individuals in mental hospitals are only marginally ill, slightly "abnormal," or suffering from senility. Nevertheless, mental illness is probably the largest medical problem today, affecting one out of ten persons and filling nearly one-half of all hospital beds in the United States. As with other social problems, mental illness must be considered in relation to a specific cultural context. Persons judged "normal" in modern western industrial society might be viewed as mentally ill in certain aboriginal societies and vice versa.

In modern western society three main types of mental

illness are generally recognized. First is the type that results from *birth and developmental defects* (such as epilepsy and some forms of mental retardation). Second is that which includes a broad range of psychological or adjustment problems resulting in abnormal behavior, which is called *neurosis*. The third type consists of those severe forms of mental illness which leave persons at times completely unable to function in their society and no longer in contact with reality. The latter is called *psychosis* and is now thought by many scientists to be chemically based. For example, some tentative evidence has led to a hypothesis that *schizophrenia* and *manic-depressive* psychoses may result from abnormally increased norepinephrine and dopamine chemical pathways within the brain. Excesses of these chemicals, according to this hypothesis, stimulate nerve transmission, and the conscious cortex becomes deluged and overwhelmed with sensory input. In *simple depression,* another form of severe psychosis, some scientists believe that the chemicals necessary for nerve stimulus transfer (norepinephrine and dopamine) are present in insufficient quantity and thus the relatively inactive brain is understimulated.

Chemical tranquilizers revolutionized the treatment of some forms of psychosis. The first, reserpine, was isolated from the *Rauwolfia* plant of India in the early 1950s. Today there are many types of tranquilizers, some of which resemble the chemical structure of norepinephrine and dopamine. They compete with them, block receptor sites on neurons, and diminish any overactivity of impulses. They range from potent tranquilizers such as chlorpromazine, used to treat psychoses, to milder types such as diazepam, used to treat anxiety.

The tranquilizers opened up new approaches to the investigation of the chemical basis of mental illness. They provided a means for calming patients in mental hospitals without putting them to sleep. Furthermore, so-called hopeless patients, while under the effects of tranquilizers, were often more amenable to treatment by psychotherapy. In some cases these drugs have dramatic effects in eliminating psychotic symptoms, and they have been a great boon to psychiatric medicine. Nevertheless, undesirable side effects, such as reduction in blood pressure, tremors, and gastric disturbances, are frequently produced.

Experiments have shown that these drugs inhibit the activities of the hypothalamus and other centers, such as the reticular activating system (RAS), and thus prevent some stimuli from reaching the conscious level of the cerebral cortex. Researchers found that azacyclonol could prevent hallucinations produced by LSD and mescaline. It is believed that hallucinogenic compounds (e.g., LSD) intensify and stimulate the brain chemical serotonin, and that tranquilizing drugs interfere with and depress the actions of serotonin in transferring stimuli across nerve synapses. Other chemicals in which the mechanism of action is not fully understood can act as antidepressants. Thus, great strides have been made in the chemical treatment of severe mental disorders.

However, most scientists recognize that the underlying causes of psychoses may lie in an interaction between physiological-chemical and psychological-environmental factors. Experimental psychoses can be induced without drugs; for example, by drastically reducing environmental stimulation. After relatively brief periods of sensory deprivation, in soundproof and darkened isolation rooms, subjects begin to have hallucinations and experience the most unpleasant psychotic reactions. Also, it appears that the isolated person's desire for stimulation makes him extremely susceptible to propaganda and techniques of "brainwashing." Some researchers believe that the family situation and relationships have an important bearing on the development of psychoses. In addition, the sociocultural environment appears to affect the incidence of psychosis. August Hollingshead and Frederick Redlich of the Yale University School of Medicine have shown that, in the United States, people in the lowest socioeconomic group were hospitalized for schizophrenia 12 times more often than those in the highest socioeconomic group. Even the ethnic culture of a group may influence the nature of the schizophrenic psychoses that develop. It may be determined that severe mental disorders are caused by chemical imbalances that are in turn somehow stimulated by a stressful environment. In any case, we are still far from a complete understanding or overall cure for mental illness.

In some patients modern techniques of brain surgery can be used to correct mental disorders. Drs. Vernon Mark and Frank Ervin of the Harvard Medical School have removed the amygdala from 13 patients who suffered from periodic violent seizures and rage; some of the patients showed marked improvement. In Japan, 56 children with serious brain damage and uncontrollable violent behavior were operated on by Doctor Keiji Sano of Tokyo University School of Medicine and all but a few became relatively calm.

Such psychosurgery holds great promise for humanely correcting some severe mental disorders, but like many of our other powerful new technologies it could, in the wrong hands, be misused for human subjugation. Many persons view it with justifiable concern and point to the necessity of proper ethical guidelines for all new biological technologies (see also Chap. 5).

Scientists hoping to isolate chemicals that could promote the regrowth of severed or damaged nerves recently discovered a chemical factor that stimulates the growth of neurons. This nerve growth factor (NGF) has a potent stimulatory effect on the regeneration of severed nerve cells. Björklund and Stenevi, of the University of Lund in Sweden, have suggested that NGF is a normally occurring physiological factor required by the nervous system for development and growth.

As understanding of the chemical nature of the brain increased, scientists began to stimulate certain localized regions of the brain with chemicals. In 1954, Alan Fisher, then at McGill University, began to experiment with chemical substances to stimulate specific brain cells. He found that injecting the male sex hormone testosterone into the brain of male rats produced in them maternal activities such as nest-building and other well-patterned motherly behavior. This indicated that specific behavior could be controlled by chemical alteration of the brain. In further experiments, Fisher implanted tiny guide shafts into the brain of rats through which he could deliver as little as 1 μg or .01 ml of chemical solution. Injection of the same hormone in an area just adjacent to the maternal activity center caused both male and female rats to engage in male sexual activity. Other scientists have confirmed that the steroid hormones can act selectively on nerve cells at specific sites in the brain. Also, Fisher found that injections of three different chemicals at the same site in the brain caused different reactions. An injection of acetylcholine stimulated the animal to drink, norepinephrine prompted him to eat, and testosterone caused him to build nests.

How Do We Remember?   More than a decade ago James McConnell of the University of Michigan startled the scientific world with a most unexpected and amazing discovery. He taught planaria (a primitive type of flatworm) to respond to a light signal, he then ground up the educated worms and fed them to untrained worms; the untrained worms acquired knowledge — that is, they responded to the light signal just as if they had been through a lengthy training program. Many researchers suddenly became interested in "chemical memory transfer." A. L. Jacobson at UCLA, followed by many other workers, extended such experiments to rats, whose brains are basically similar to that of man. They found that when brain extracts from trained rats were injected into untrained rats the untrained rats acquired the memory of the "smart" donor rat. Many neurophysiologists, however, still doubt at this time the conclusiveness of memory transfer experiments.

During this same period, a group of researchers (D. Krech, M. R. Rosenzweig, E. L. Bennett, and others) at the University of California were investigating the effects of environment on brain development. They divided rats into two groups; both received identical sanitary care, food, and water, but their psychological environments differed greatly. The first group was placed in an *enriched environment* cage, provided with many rat toys and other rats to play with. They were taught to run mazes and were given loving human care. The rats in the second group were placed in a *deprived environment* of isolated, barren cages devoid of toys, fellow rats, and extensive human contact. After 80 days they found the brains from the enriched environment rats had a heavier cerebral cortex, better blood supply, larger brain cells, and increased activity of two important brain enzymes (acetylcholinesterase and cholinesterase) compared to the brains of the deprived rats.

Researchers are now attempting to identify a chemical basis for learning and memory. Bernard Agranoff of the University of Michigan has found that goldfish can learn to swim to one side of a tank to avoid an electric shock and remember this response for several days. If the fish are injected with chemicals that inhibit protein or nuclear RNA (ribonucleic acid) synthesis just before or just after training, their memory for shock-avoidance behavior is not immediately impaired. Such *short-term* memory does not involve synthesis of new proteins; however, when tested a day or two later for *long-term memory*, goldfish showed almost no memory of their previous training. Such findings indicate that long-term memory may depend upon the synthesis of certain chemicals (perhaps new proteins) by brain cells. Other researchers find that injecting rats with certain chemical compounds that stimulate the central nervous system leads to a significant increase in memory ability.

In December, 1970 Dr. George Ungar of the Baylor College of Medicine announced the isolation of a psychoactive chemical that may play an important role in the chemistry of memory and learning. Rats (which normally prefer darkness) were trained by operant conditioning (see p. 124) to fear the dark. From the brains of such dark-fearing rats he isolated a chemical compound later to be named *scotophobin*. When as little as 0.1 μg of scotophobin was injected into an unconditioned rat, it changed its normal preference for dark into an acute fear of darkness. Learning apparently had been transferred chemically from the brain of one mammal to another.

Scotophobin has now been synthesized in the laboratory. Some psychochemists believe that such memory chemicals represent the messages of a "learning and memory" code similar in some respects to the coded messenger chemicals of DNA (see Chap. 5). Such findings give hope for the possibility of a chemical treatment for mental retardation or faulty memory of the aged. Futurists now predict (only partly in jest) that not-too-distant day when schools and books become outmoded and learning is transferred by injection from professor to student. If and when that day arrives we shall be faced with the responsibility of protecting our freedom to choose *what* we learn.

ALTERED STATES OF CONSCIOUSNESS

> Our normal waking consciousness . . . is but one spe-
> cial type of consciousness, whilst all about it, parted from it
> by the filmiest of screens, there lie potential forms of con-
> sciousness entirely different. . . . No account of the universe
> in its totality can be final which leaves other forms of con-
> sciousness quite disregarded.

William James, 1929

Certain physiological, psychological, or pharmacological manipulations can produce mental states significantly different from recognized normal patterns characteristic of alert waking consciousness. These altered patterns are called "altered states of consciousness" (ASC).

Many psychologists now believe that an optimal range of incoming sensory stimuli (or a proper balance between excitatory and inhibitory stimuli) is necessary for maintaining "normal" waking consciousness, and that levels of stimulation either above or below this range may produce ASCs. Lowering of sensory stimulation may result from absolute reduction of sensory input or motor activity, or constant exposure to repetitive monotonous stimulation (as may result from solitary confinement, extreme boredom, experimental sensory deprivation, or bodily immobilization) or from the "passive state of mind" achieved through meditation. On the other hand, heightening of sensory stimuli may be achieved through increased alertness, mental involvement, and concentration, or from excitatory mental states resulting primarily from sensory overload or bombardment, which may or may not be accompanied by strenuous physical activity (e.g., "third degree" brainwashing techniques, trances produced during spiritual revivalistic meetings, and perhaps schizophrenic reactions). Also, disturbances in body chemistry (such as high or low blood sugar, dehydration), as well as deliberate administration of anesthetic, psychedelic, narcotic, sedative, or stimulant drugs can result in ASCs. Hypnosis and sleep dreaming may also be considered ASCs. These are only some of the many ways in which consciousness can be altered. We shall consider in more detail only two of the most widely practiced ways of achieving ASCs—psychoactive drugs and meditation.

Should We Alter Our Brains Chemically?

This may be a moot question—many Americans are already altering the chemistry of their brain in one way or another. For example, in 1972 the author anonymously surveyed

TABLE 4–1 Use of Mind-Altering Drugs*

	N.R.	Frequently	Occasionally	Once	Never	Total used at least once (% of responses)
Caffeine (coffee, tea)	–	18	12	1	4	31 (89%)
Alcohol (any form)	2	11	20	1	1	32 (97%)
Tranquilizers	2	0	2	4	27	6 (18%)
Stimulants (benzedrine, etc.)	2	0	6	5	22	11 (33%)
Hallucinogens (LSD, mescaline, psilocybin, etc.)	1	1	5	5	23	11 (32%)
Cannabis (marijuana, hashish, etc.)	1	7	13	2	12	22 (65%)
Nicotine (tobacco)	11	8	5	2	9	15 (62%)

*A total of 35 students in a suburban community college were asked to reply confidentially to a questionnaire on drug use and to respond to "I have used the following:". (N.R. = no response.)

35 suburban college students (age 18 to 45) with regard to their use of the mind altering drugs caffeine, alcohol, tranquilizers, stimulants, hallucinogens, and marijuana. Although the sample is small and should not be taken to represent any overall trend, it is of interest that of the 35: (1) all had used one or more of the drugs listed at least once, (2) more had used marijuana at least once (65 per cent) than tobacco (62 per cent), (3) more than half (57 per cent) were frequent or occasional users of marijuana, and (4) almost one-third (32 per cent) had used hallucinogens (LSD, etc.) at least once (Table 4–1).

Probably a more realistic question would be not whether we should use such drugs but rather what are the rewards and possible ill effects of drugs already in use. As an example let us explore what we currently know about the hallucinogenic drugs.

What Are the Effects of Marijuana? Marijuana, like many mind-altering drugs, has been used for thousands of years. It was recorded in a volume on Chinese medicine in 2737 BC; it grows as a common weed throughout many areas of the world including the United States, and is used by over 200 million people. How many people in the United States use marijuana? Few widespread studies exist, but a 1969 study in San Francisco indicated that in the 18- to 24-year-old age group, one-half of the men and one-third of the women surveyed had used marijuana at least once. Furthermore, the proportion who used marijuana was as great among nonstudents as among students.

In the United States the ground leaves of the marijuana plant *Cannabis sativa* are generally smoked as a cigarette or in a pipe. Smoking or ingesting marijuana or its stronger derivative, hashish, generally produces a mild euphoria and enhances the

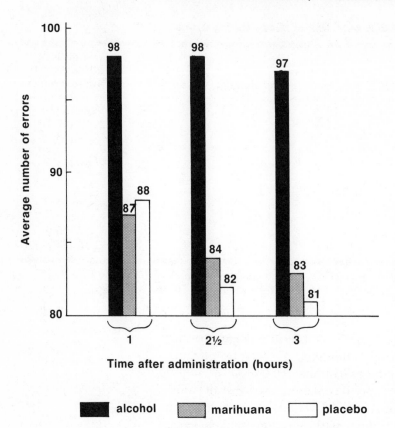

Figure 4–8 Comparative study of the effects of marijuana and alcohol on simulated driving performance was conducted by the Bureau of Motor Vehicles of the state of Washington. The graph shows average number of errors on tests administered at the stages after treatment with alcohol (a), marijuana (b), and a placebo (c). In general it was found that marijuana causes significantly less impairment of driving ability than alcohol does. (Adapted from Grinspoon, L.: Marijuana. Sci. Am. 221(6):17–25, 1969.)

sensory perception of colors and sounds. In stronger doses, mild hallucinations may be produced. Another effect of the drug is the lengthening of apparent time; under the influence of marijuana 10 minutes may seem like an hour. The primary (but probably not the only) active chemical of *Cannabis sativa* is tetrahydrocannabinol (THC), a nonaddictive drug. There is no convincing evidence that it causes genetic damage; however, THC remains in the body for a long period of time. The long-term effects of the drug are not completely understood. Marijuana, although producing some of the same effects as the stronger hallucinogenic drugs (LSD, mescaline, and psilocybin), is definitely less potent and consciousness-altering. Few physical symptoms accompany the use of marijuana other than an increase in pulse rate, a slight increase in thirst, and a striking increase in hunger and appreciation of food. There is no indication that marijuana incites people to aggressive or violent behavior. Marijuana causes significantly less impairment in motor abilities and coordination (e.g., those needed in driving an automobile) than do other drugs such as alcohol (Fig. 4–8). With the increasing use of marijuana by western society, scientific research into its effects has increased. The availability of THC in synthetic form has led to experiments in which the dose can be measured with greater precision.

Leo Hollister reviewed the results of research on marijuana up to 1971. He found that in controlled experiments (1) no changes in pupil size, respiratory rate, or reflexes occurred; (2) blood pressure tended to fall slightly or remain unchanged, and pulse rate increased; (3) perceptual and psychic changes included pronounced euphoria, alteration of the time sense, and less discrimination in hearing; (4) vision was apparently sharper but with some visual distortions; (5) there was some difficulty in concentrating and thinking, and dreamlike states were prominent; (6) on "self-reporting mood scales" subjects were more friendly after initially taking the drug but less so with the passage of time; (7) the subjects were less aggressive, thought less clearly, and especially after three hours, became sleepy; (8) subjects maintained their accuracy in answering arithmetic tests but their performance was slowed; (9) a drawing test showed reduced accuracy but no slowing of performance; (10) the accuracy on mathematical serial addition tests was impaired but the drug had no effect on the ability to count backwards, to say the alphabet backwards, or to repeat digits forward or backward; (11) tests comparing the effects of extracts of marijuana in various doses indicated that whereas long-term memory was maintained, short-term memory was impaired; (12) tests indicated an impairment of the more complex task — goal-directed behavior in which successive subtractions and additions had to be made to reach a specified end number. Disturbance of this type of goal-oriented behavior has been termed "temporal disintegration"; (13) declines in reading comprehension under the influence of oral doses of marijuana extract were explained as a loss of the selective attention span and of systematic thinking.

Answers to questionnaires by users of marijuana confirm the symptoms reported by laboratory studies — i.e., floating sensations, depersonalization, relaxation, perceptual changes (visual, auditory, and tactile), subjective slowing of time, flight of ideas, difficulty in concentrating and loss of attention, loss of immediate memory, euphoria, silliness, sleepiness, increased perception and insight, and increased sexual desire, performance, and enjoyment. Many of the adverse reactions that have been attributed to marijuana may be due to the adulteration of the drug with other substances when it is sold illegally.

Is marijuana preferable to alcohol as a social drug? Past experience indicates that rather than replacing other drugs, new drugs are used in addition to those already in existence. Can users of marijuana develop a dependence? There is no evidence of physical dependence or addiction to marijuana but, being an animal of habit, man can develop psychological dependence to many drugs (e.g., tobacco and coffee).

Does smoking marijuana lead to the use of stronger drugs

and addictive narcotics? Most users of addictive opiate drugs such as heroin also use or have used marijuana frequently. However, from such a correlation we should not be led to the conclusion that narcotic addiction results from marijuana use. It is also true that most narcotic addicts have eaten potatoes — but eating potatoes is most probably not the cause of their narcotic addiction. Some correlation between marijuana use and narcotic addiction may result from present laws that force marijuana users to deal with black market individuals who may also push hard narcotics. Nevertheless, there is no evidence that opiate addiction *results from* prior use of marijuana.

One possible danger in the chronic use of marijuana may lie in its subtle effects on personality. Some studies have indicated that prolonged habitual use of marijuana may result in loss of desire to work, loss of motivation, loss of intellectual function, and possible interference with goal-oriented behavior. However, other studies conclude that there is no real link between marijuana and the "amotivational syndrome." The effects of long-term and widespread chronic use of marijuana on individual personality and on society will require continuing study and analysis.

What Are the Effects of the Potent Hallucinogenic Drugs?
Hallucinogenic or psychoactive drugs, primarily derived from plants, have been used by man for thousands of years in magical, medical, and religious practices. Among the 400,000 to 800,000 known species of plants, about 60 are presently known to contain psychoactive (mind-altering) compounds and probably 20 of these are considered important. However, estimates indicate that as many as 5000 higher alkaloid-containing plants may turn out to be hallucinogenic.

Within the last decade three types of hallucinogen have grown in widespread use throughout the western world. These are mescaline, psilocybin, and lysergic acid diethylamide (LSD) (Fig. 4–9). Mescaline is a psychoactive alkaloid chemical derived from the peyote cactus *Lophophora williamsii*, which grows in the deserts of central and northern Mexico. It was used widely in pre-Columbian religious practices of Aztec and other Mexican Indians. Late in the nineteenth century its use spread among the Plains Indians of the United States and it became incorporated into a religious dogma that fused with Christianity. In 1918 the Native American Church was formed with peyote ceremonies as part of its religious practice — it now has more than 250,000 followers.

Psilocybin comes from the mushrooms *Stropharia cubensis* and *Psilocybe mexicana,* among others. Along with mescaline, it played an important part in religious rites among the pre-Columbian Indians. The ancient Aztecs called these "sacred" mushrooms *teonanacatyl* ("food of the gods"). It is still used

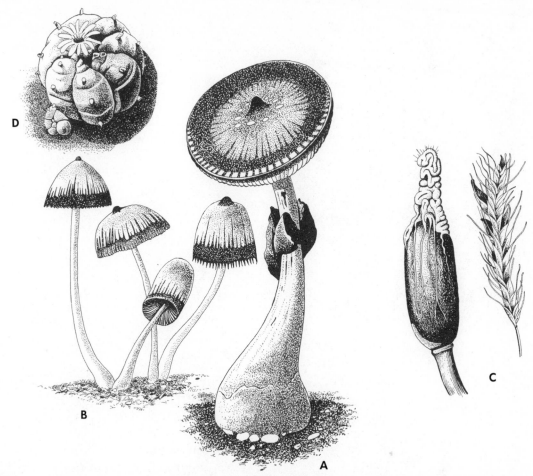

Figure 4–9 Natural sources of the major hallucinogens are depicted. Psilocybin comes from the mushrooms *Stropharia cubensis* (*A*) and *Psilocybe mexicana* (*B*). LSD is synthesized from an alkaloid in ergot (*Claviceps purpurea*), a fungus that grows on cereal grains; an ergot-infested rye seed head is shown (*C*) together with a larger scale drawing of the ergot fungus. Mescaline is from the peyote cactus Lophophora williamsii (*D*). (Adapted from Barron, F., et al.: The hallucinogenic drugs. Sci. Am. *210*(4):29–37, 1964.)

today by tribes in southern Mexico. Studies during the 1950s indicated that there are more than 24 species of mushrooms utilized for their hallucinogenic properties by six or more tribes of Mexican Indians. The effects of hallucinogenic mushrooms were alluded to by Lewis Carroll in his story of *Alice in Wonderland* (Fig. 4–10).

The chemical structures of psilocybin and LSD are very similar to the neurochemical serotonin involved in transferring nerve impulses (see earlier in this chapter). The chemical structure of mescaline closely resembles the two neurochemicals epinephrine and norepinephrine (Fig. 4–11). Since most of the controlled laboratory studies have been done on LSD, we will review knowledge concerning this drug, bearing in mind that the subjective effects and psychoactivity of all three hallucinogens are similar.

The subjective effects of a hallucinatory "trip" are widely documented and well known. These include a dreamy state ac-

Figure 4–10 (Taken from The Annotated Alice by Lewis Carroll with an introduction and notes by Martin Gardner © MCMLX by Martin Gardner. Used by permission of Clarkson N. Potter, Inc.)

companied by feelings of changes in body image. For example, the limbs may feel peculiar (short, long, light) and the skin may feel sensitive or numb. The subject may have complete awareness and feel somewhat detached and outside of his own body. Changes in visual perception include intensely bright and pure color sensations and hallucinations. As with marijuana, time appears to pass more slowly. There is no indication of amnesia (loss of memory) after a "trip." Laboratory studies indicate that subjects develop a short-term tolerance to LSD; that is, in subjects taking the drug several times day after day, progressively larger doses are required to produce the same effect. This increased dosage requirement is lost rapidly when the drug is no longer taken. "Bad trips" and paranoid reactions under the influence of LSD are rare (less than 1 per cent), but are most likely to occur among people who show slight tendencies towards paranoia on pretests. Subjects who show the most intense and pleasant reactions to the drug are shown by pretests to be high in aesthetic sensitivity and imagination. They show a preference for an unstructured and inward-turning lifestyle, and have an orientation to ideas; they are less aggressive, competitive, and dogmatic as well as less conforming. In short, studies show that personality variables may play an important role in determining both the overall intensity of the reaction to LSD and the degree to which the subject becomes anxious, hostile, or paranoid under the influence of the drug.

Some physical or physiological changes can be noted under

the influence of LSD. These include dilation of the pupils, increase in heart rate, elevation of body temperature, piloerection (raising of the body hair), and hyperglycemia (high blood sugar). These reactions may result from a stimulation of the sympathetic nervous system similar to that which occurs when a person is excited or under stress. Usually no change in the wave pattern of the electroencephalogram (brain waves) occurs.

Can LSD produce chromosome breakages or birth defects? In 1967 experiments indicated that when LSD was added to human white blood cells in a test tube, chromosome breakages were produced. Other studies also indicated the possibility of human chromosome damage resulting from LSD. Researchers found that blood samples from LSD users contained white blood cells with 19 per cent chromosome breakage compared to

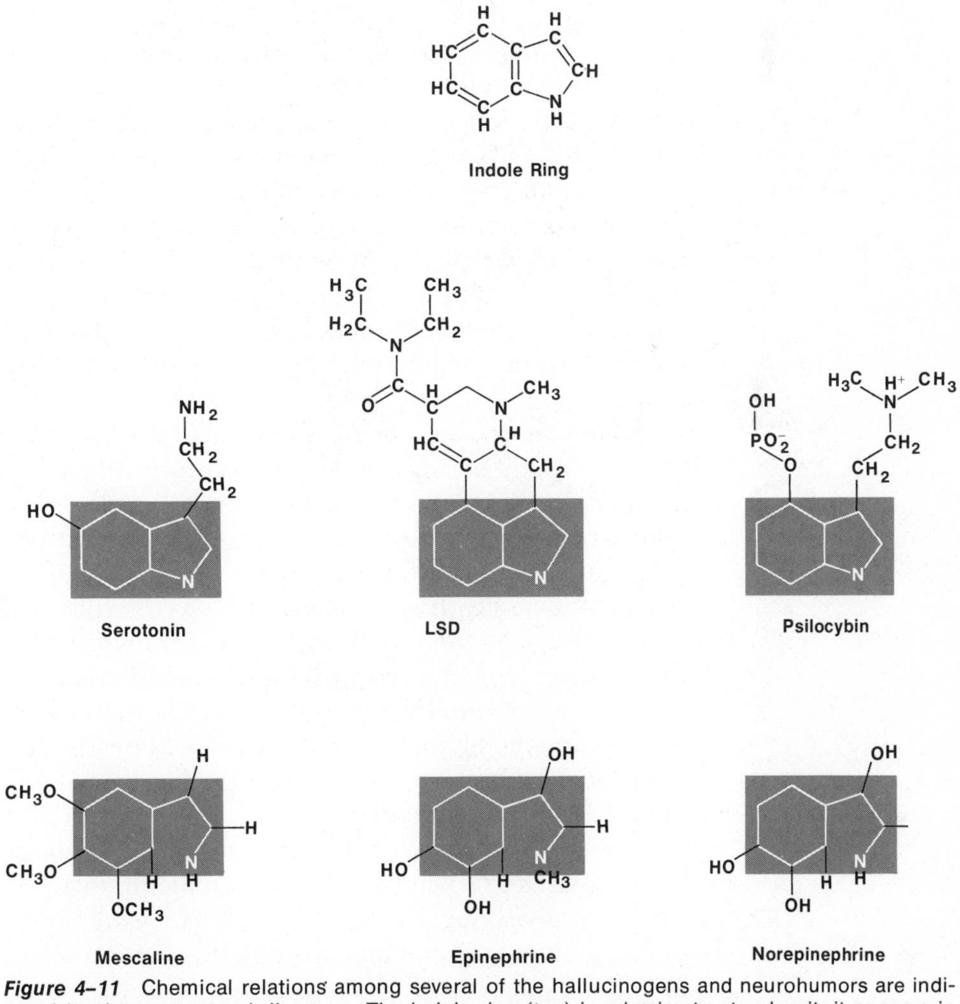

Indole Ring

Serotonin   LSD   Psilocybin

Mescaline   Epinephrine   Norepinephrine

Figure 4-11 Chemical relations among several of the hallucinogens and neurohumors are indicated by these structural diagrams. The indole ring (top) is a basic structural unit; it appears in serotonin, LSD, and psilocybin. Mescaline does not have an indole ring, but can be represented so as to suggest its relation to the ring. The close relation between mescaline and the two catecholamines epinephrine and norepinephrine is also apparent here. (Adapted from Barron, F., et al.: The hallucinogenic drugs. Sci. Am. 210(4):29–37, 1964.)

only 9 per cent for nonusers. However, the total number of chromosome breakages was not correlated with the total dosage of LSD and the question remained whether the results could have been produced by other drugs which these persons were using in addition to LSD. Another study found an increased incidence of chromosome abnormalities in the reproductive cells of mice that were injected with LSD, but the dose used in the study was six to eight times the typical human dose. A study of mental patients who had been treated with daily doses of LSD for periods ranging between 5½ and 35 months and who had received no other drugs, revealed no increased incidence of chromosome damage; however, the latter study was made 20 to 24 months after exposure to LSD had stopped. Other studies have found no chromosome abnormalities produced by LSD, and some scientists criticize previous workers for failure to measure chromosome breakages prior to LSD use. High doses of LSD injected into pregnant mice early in pregnancy were found to cause severe birth defects. In another study, however, lower doses had no effect on mice offspring and several other workers have since found that pregnant rabbits or mice injected with LSD show no increased incidence of birth defects.

After an extensive review of more than 30 studies of LSD performed prior to 1971, Norman Dishotsky and other scientists concluded that chromosome damage, when found, was related to drug abuse in general and to adulterated illicit LSD in particular. Pure LSD ingested in moderate dosages did not increase the occurrence of chromosome damage, birth defects, or cancer.

Can Consciousness Be Altered without the Use of Drugs?
One means of altering mind consciousness without drugs has been practiced for thousands of years, especially in Asia. This is meditation. Various techniques of meditation (and there are many) attempt to produce a state of contemplation or deep passivity combined with awareness in which "the conscious exercise of attention leads to a spontaneous flow of experience to which the person becomes a receptive onlooker" (Tart, 1969). Meditation is growing rapidly in popularity in the western world. More than 300,000 people in the United States and elsewhere have received instruction in one of the many types known as "transcendental meditation."

Studies made during the 1950s and 1960s indicate that meditation as practiced by yoga or Zen meditators produces a variety of physiological effects including a reduction in the rate of metabolism, presumably through unconscious control of the involuntary autonomic nervous system. Indications were also found of an increase in certain types of electrical waves given off by the brain (alpha waves) and of an increase in the electrical resistance of the skin. Both of these findings are thought to

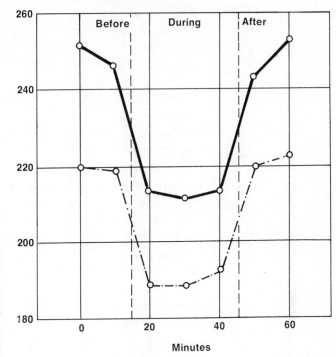

Figure 4-12 Effect of meditation on the subjects' oxygen consumption (solid line) and carbon dioxide elimination (broken line) was recorded in 20 and 15 cases respectively. During meditation, both rates decreased markedly (dash area). Consumption and elimination returned to the premeditation level soon after the subjects stopped meditating. (Adapted from Wallace, R. K., and H. Benson: The physiology of meditation. Sci. Am. 226(2):84–90, 1972.)

contribute to some extent to a decrease in the level of anxiety and an increase in the relaxed state.

Keith Wallace and Herbert Benson made a systematic study of the physiological effects or "correlates" of meditation by studying 36 subjects who were experienced in transcendental meditation. Their results indicate that the meditative state can be described as "wakeful, hypometabolic." Such a state is characterized by: (1) reductions in oxygen consumption, carbon dioxide elimination, and the rate and volume of respiration (Fig. 4–12); (2) a slight increase in the acidity of the arterial blood; (3) a marked decrease in blood lactate level; (4) a slowing of the heartbeat; (5) a considerable increase in skin resistance; and (6) an electroencephalogram pattern of intensification of slow alpha brain waves with occasional beta-wave activity.

The meditative state appears to be an integrated pattern of responses mediated by the central nervous system in which several characteristic changes take place. Another type of patterned response is the "fight or flight" or defense alarm pattern of reaction to stressful situations. There, the sympathetic nervous system is mobilized to meet a sudden challenge and produces a marked increase in blood pressure, heartbeat, blood flow to the muscles, and oxygen consumption. The "hypometabolic state" produced by meditation appears to be the opposite in many respects to the "fight or flight" reaction to stress.

Critics feel that meditation is essentially a withdrawn state in which practitioners could lose objective control over the necessary outward planning of ordinary everyday activities and come to value "internal" at the expense of "external" experience. Increasingly, however, scientists and laymen are recognizing important benefits derived from meditation. The meditative state is essentially a very relaxed psychological state and could have beneficial effects especially for people suffering from anxiety neuroses or hypertension. Meditators frequently report calm, improved sleep, and improved energy for constructive work. The impact of rapidly changing contemporary technology may have harmful effects on man's psychological abilities to cope. Although used most widely by educated middle and upper classes, meditation may offer some promise for psychological health in a technological world.

Are the Psychotic, Drug-Induced, and Meditative Alterations in Consciousness Related? In an intriguing paper, Roland Fischer of the Ohio State University College of Medicine places the psychotic, hallucinatory-drug, and meditative states on a continuum scale of hallucination-perception. According to Fischer, beginning with the normal everyday state of consciousness called "I" we can draft a map of "innerspace"—first along a perception-hallucination continuum of increasing arousal (i.e., hyperarousal) of the sympathetic nervous system which includes creative, psychotic, and ecstatic experiences and secondly along a perception-meditation continuum of decreasing arousal (i.e., hypoarousal). Both states are marked by a gradual turning inward toward the mental dimension at the expense of the physical dimension (Fig. 4–13).

Hallucinations are described as "experiences of intense sensation that cannot be verified through voluntary motor activity." Some objective measurements appear to relate to these altered states of consciousness. Besides the physiological changes accompanying meditation (discussed earlier in this chapter), the perception-meditation continuum is characterized by EEG brain waves of progressively lower frequencies, as well as by a *slowing* of the rapid eye movements. On the other hand, the frequency of rapid eye movements *increases* five to eight times in response to the hyperarousal induced by moderate doses of hallucinogenic drugs and is also present without drugs in acute schizophrenics. Also, with increasing hyperarousal, subjects need fewer molecules of a substance to elicit a noticeable taste difference such as sweetness or bitterness. In either direction along the continuum of hyperarousal or hypoarousal the subject experiences a contraction of space and time until reaching the "self." Here space and time, as well as the difference between the observer and the observed or between subject and object, disappear and the subject becomes "one" with the universe. At this point there is no longer an ability to verify sense experience through voluntary motor activity,

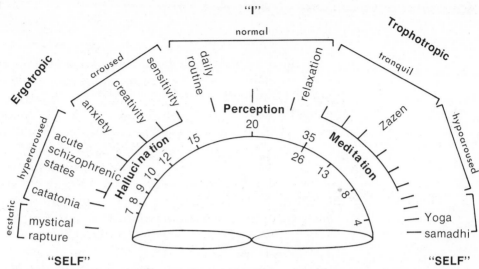

Figure 4–13 Varieties of conscious states mapped on a perception-hallucination continuum of increasing "ergotropic" arousal (left) and a perception.meditation continuum of increasing "trophotropic" arousal (right). These levels of hyper- and hypoarousal are interpreted by man as normal, creative, psychotic, and ecstatic states (left) and Zazen and samadhi (right). The loop connecting ecstasy and samadhi represents the rebound from ecstasy to samadhi, which is observed in response to intense ergotropic excitation. The numbers 35 to 7 on the perception-hallucination continuum are Goldstein's coefficient of variation (46), specifying the decrease in variability of the EEG amplitude with increasing ergotropic arousal. The numbers 26 to 4 on the perception-meditation continuum, on the other hand, refer to those beta, alpha, and theta EEG waves (measured in hertz) that predominate during, but are not specific to, these states. (From Fischer, R.: A cartography of the ecstatic and meditative states. Science *174*(4012): 897–904, 1971. Copyright 1971 by the American Association for the Advancement of Science.)

nor is there any communication between the "I" of the everyday world and the "self" of the ecstatic state. At moderate levels of hyper- and hypoarousal, communication between the "I" and the "self" is still possible. Fischer believes that this communication is actually the creative source of art, science, literature, and religion.

This ecstasy or mystical experience of the oneness of everything results from an increasing integration of the higher brain centers of the mind (cerebrum) with the lower centers of the brain in the subcortex. To summarize, in Fischer's own words:

> During the "I" state of daily routine the outside world is experienced as separate from oneself, and this may be a reflection of the greater freedom (that is, separateness or independence) of cortical interpretation from subcortical activity. With increasing [hyper- or hypo-] arousal, however, this separateness gradually disappears, apparently because in the "Self"-state of ecstasy and samadhi, cortical and subcortical activity are indistinguishably integrated. This unity is reflected in the experience of Oneness with everything, a Oneness with the universe that is one-self.

HOW SHALL WE SHAPE HUMAN BEHAVIOR?

Behavioral psychologists believe that behavior departing widely from accepted social norms (neuroses) results primarily not from some organic or chemical disorder but from learned behavior that is maladaptive. There can be little doubt that behavior is learned or that it can be taught. Many laboratory experiments with animals show that behavior can be "shaped" by *operant conditioning*. In such conditioning, undesirable responses are punished by "negative reinforcement" (e.g., by administration of an electric shock). "Positive reinforcement," in which the desired behavior is rewarded, for example, with food, may be even more effective in shaping a desired behavior. In laboratory experiments animals learn in stepwise progression that pressing a lever will reward them with food only after they have performed certain tasks (designed by the experimenter). Through step-by-step rewarding or "shaping," the desired behavior can be produced. Rats can be taught to negotiate a complex maze, and pigeons can be taught such complex behavior as playing ping-pong (Figs. 4–14 and 4–15).

Can these principles of conditioning and behavior control developed in laboratory animals be applied to human behavior? B. F. Skinner, psychology professor at Harvard University and probably the most influential of modern behavioral psychologists, said in 1925: "Give me a dozen healthy infants, and I'll guarantee to take any one at random and train him to become any type of specialist I might select—doctor, lawyer, even beggarman and thief, regardless of his talents, penchants, tendencies, or abilities." Indeed, there is increasing evidence that, like other animals, man reacts to behavioral conditioning. Psychologist Albert Bandura of Stanford University believes that positive reinforcement conditioning can be an important procedure in the treatment of abnormal human behavior. Hyperaggressive children, when taught that aggressive reactions resulted in unpleasant consequences, and that cooperative reactions resulted in rewards, showed a lasting decrease in aggressive, domineering behavior compared with control groups. Fears (phobias) and inhibitions can be eliminated by having the fearful person observe a graduated sequence of activities beginning with those that are easily tolerated and progressing to more fearful situations. For example, Gordon L. Paul at the University of Illinois showed that conditioning techniques were extremely effective in reducing the fears of college students who suffered from extreme anxiety about public speaking.

Behavioral training has been used in the treatment of juve-

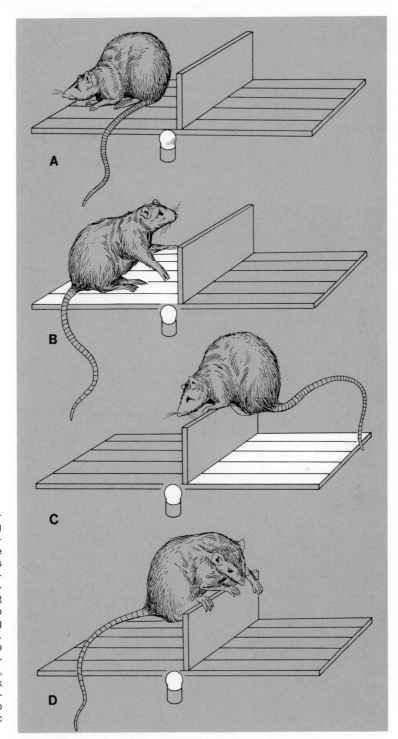

Figure 4–14 In a typical negative reinforcement conditioning experiment a "naive" (untrained) rat does not behave (move) when a red light comes on. When the red light is immediately followed by punishment (an electric shock), it soon learns by trial and error to avoid the shock by jumping over to a nonelectrified box. Soon the rat learns to jump into the nonelectrified box whenever the red light comes on, even before any electric shock is received. He is now conditioned to behave in response to a light rather than an electric shock.

nile delinquency. In an institute for delinquents, youths were rewarded with points for desirable behavior such as picking up books or going to lectures; these points could then be exchanged for privileges such as better food, a private room, or

Figure 4–15   "Boy, have I got this guy conditioned! Every time I press the bar down, he drops in a piece of food." (Reprinted by permission of *Jester* of Columbia University.)

television viewing time. They showed marked improvements in desirable behavior, and among those who received treatment the rate of return to the institute dropped from 85 to 25 per cent.

Such behavioral modification may have important consequences for the future of mankind. Faced with our modern technologies of mass destruction and with the skyrocketing increase in the rate of change itself, human behavior will have to change if man is to survive. Yet the control of human behavior is certainly one of the most frightening prospects for the future. Who will decide what behavior is normal and what is abnormal, what is desirable and what is undesirable? Must we look forward to a 1984 Orwellian state in which nonconformists and "abnormals" are brought into "learning centers" for reconditioning to behavior deemed desirable by the authorities? As Harvard social psychologist Herbert C. Kelman has warned:

For those of us who hold the enhancement of man's freedom of choice as a fundamental value, any manipulation of the behavior of others constitutes a violation of their essential humanity, regardless of the "goodness" of the cause that this manipulation is designed to serve.

REFERENCES

Andrews, G., and S. Vinkenoog, eds.: The Book of Grass: An Anthology of Indian Hemp. New York, Grove Press, 1967 (paperback). (Collected essays and stories dealing with *Cannabis sativa*.)

Asimov, I.: The Human Brain: Its Capacities and Functions. New York, New American Library (Signet), 1963 (paperback). (Readable account of hormones, glands, nervous system, brain, senses, etc.)

Bandura, A.: Behavioral psychotherapy. Sci. Am. *216*(3):78–86, 1967. (Abnormal behavior is a problem of "social learning.")

*Barber, T. X.: LSD, Marihuana, Yoga, and Hypnosis. Chicago, Aldine Publishing Co., 1970. (A review of research into physiology and effects of altered states of consciousness.)

*Barron, F., M. E. Jarvik, and S. Bunnell, Jr.: The hallucinogenic drugs. Sci. Am. *210*:29–37, 1964. (The nature, history, and effect of hallucinogens.)

Björklund, A., and U. Stenevi: Nerve growth factor: Stimulation of regenerative growth of central noradrenergic neurons. Science *175*:1251–1253, 1972. (A technical research report.)

*The Brain. Life, 1971. I (Oct. 1), II (Oct. 22), III & IV (Nov. 26). (4 part series. I — The structure of the brain and sense organs, II — The neuron, III — How the brain controls consciousness and behavior, IV — The chemistry of madness and drugs.)

†Crichton, M.: The Terminal Man. New York, Alfred A. Knopf, Inc., 1972. (A novel about surgical-electronic mind control. Includes a good bibliography.)

Evans, W. O., and N. S. Kline, eds.: The Psychopharmacology of the Normal Human. Springfield, Illinois, Charles C Thomas, 1969. (A review of research on the nature of normality, and the effects of drugs.)

Fisher, A. E.: Chemical stimulation of the brain. Sci. Am. *210*(6):60–68, 1964. (Injection of chemicals into localized areas of animal brains releases drives such as maternal behavior, hunger, and thirst.)

Fischer, R.: A cartography of the ecstatic and meditative states. Science *174*(4012):897–904, 1971. (An integrated theory relating the psychotic, hallucinogenic, creative, normal, meditative, and ecstatic states of consciousness.)

Grinspoon, L.: Marihuana. Sci. Am. *221*(6):17–25, December 1969. (The history of its use, effects, and controversy surrounding effects on sexuality.)

Himwich, H. E.: The new psychiatric drugs. Sci. Am. *193*(4):80–86, 1955. (The discovery of and early work with tranquilizers.)

Hollister, L. E.: Marihuana in man: three years later. Science *172*:21–29, 1971. (A review of research into the effects on man.)

Huxley, A.: The Doors of Perception. New York, Harper & Row, 1954. (A personal account of experiences with the hallucinogen mescaline.)

Jackston, D. D.: Schizophrenia. Sci. Am. *207*(2):65–74, 1962. (Describes theories of schizophrenia and favors a psychological rather than physiological explanation.)

Luria, A. R.: The functional organization of the brain. Sci. Am. *222*(3):66–72, 1970. (Sensory and motor functions are well localized but complex functions such as speech and writing remain obscure.)

Manheimer, D. I., G. D. Mellinger, and M. B. Balter: Marijuana use among urban adults. Science *166*:1544–1545, 1969. (Survey correlates use with factors such as sex, income, and religious affiliation.)

Mind: from memory pills to electronic pleasures beyond sex. Time, April 19, 1971, pp. 45–52. (A general review of mind alteration and its possible problems for society.)

Medall, W., and A. Lansing, eds.: Drugs. New York, Time-Life Books, 1969. (A well-illustrated general review of mind-altering as well as other drugs.)

Morgan, C. T.: Physiological Psychology. New York, McGraw-Hill, 1965. (A textbook of sensory and motor functions, etc.)

Olds, J.: Pleasure centers in the brain. Sci. Am. *195*(4):105–116, 1956. (Describes the discovery of the pleasure center.)

Schultes, R. E.: Hallucinogens of plant origin. Science *163*:245–254, 1969. (Distribution of hallucinogenic compounds in plants and their early use by man.)

Shafer, R. P., ed.: Marihuana: A Signal of Misunderstanding. New York, The New American Library (Signet), 1972 (paperback). (The official report of the National Commission on Marihuana and Drug Abuse.)

Skinner's utopia: panacea, or path to hell? Time, September 20, 1971, pp. 47–53. (A general review of behavioral psychology, especially that of B. F. Skinner.)

Tart, C. T., ed.: Altered States of Consciousness. New York, John Wiley and Sons, Inc., 1969.

Wallace, R. K., and H. Benson: The physiology of meditation. Sci. Am. *226*(2):84–90, 1972. (Physiological changes in metabolism accompany meditation.)

*Whalen, R. E., ed.: Psychobiology. San Francisco, W. H. Freeman & Co., 1967. (Collected readings from Scientific American.)

*Wilson, J. R.: The Mind. New York, Time-Life Books, 1969. (A well illustrated general account of the human brain, intelligence, psychoses, and mind alteration.)

Wooldridge, D. E.: The Machinery of the Brain. New York, McGraw-Hill, 1963. (The structure and function of the brain.)

----

*Recommended further reading
†Fiction

PROBLEMS

1. Could diet control chemical imbalances such as those that cause schizophrenia and other mental diseases?

2. What is bioelectronic feedback training? How is it used, and is it dangerous to one's mental health?

3. Why is marijuana illegal? Can it harm anyone?

4. If marijuana were legalized what would be the effect on the community? Rebellion by older people? A higher crime rate?

5. What effect are drugs such as hallucinogens and marijuana likely to have on society in the years ahead?

6. What effects does "speed" (amphetamine) have on the human body?

7. Is not mind manipulation similar to manipulating the environment, and will not the result be the same?

8. Can you locate any references or articles discussing the possible effects on the brain from long-term use of drugs such as LSD, marijuana, and heroin?

9. Although the evolution of the human brain undoubtedly improved man's chances of survival in competition with other animals and nature, some have said that the brain represents such a degree of specialization that it could be the ultimate cause of human extinction. Explain how this could be.

10. What is an "electrochemical potential"? How does it operate in the human nervous system?

BRAINSTORM

It is the year 2001. The growing popularity of electrical brain stimulation (ESB), bioelectric feedback alpha pacers, mind-altering drugs and other methods of producing altered states of consciousness (ASCs) have come into widespread usage. "Elads" (electrical addicts) are becoming common. Concern has arisen among some people that widespread altering of consciousness poses a threat to contemporary society.

A legislative committee (with you as chairman) is formed in

order to determine what laws (if any) should be instituted to regulate such activities. What action will you recommend?

CONSIDER:
  (1) Considerable differences exist between such drugs as marijuana and the potent hallucinogenic drugs such as LSD.
  (2) Possible actions might range somewhat as follows:

no laws	⟷	some regulations for specific activities	⟷	severe penalties
These activities do not harm others, but affect only the individual involved			These activities could be dangerous to society	

5

THE NEW BIOLOGY — PROMISE OR PERIL?

O brave new world that has such people in it.

**William Shakespeare
The Tempest**

Might twenty-first century man be more perfect and less free?

**Walter Cronkite in CBS film
Miracle of the Mind**

WHAT IS A GENE?

One of the greatest scientific accomplishments of all time occurred during the last two decades when scientists in biochemistry, genetics, microbiology, and x-ray crystallography together produced major breakthroughs in our understanding of the structure and function of the basic unit of inheritance — the gene. Intensive work by many investigators, including several Nobel prize winners, led to an understanding of how genetic information is carried within the cell and transmitted to future generations. We present here only a brief outline of this story; the reader is referred to references at the end of this chapter for further explanations.

By the late nineteenth century, scientists realized that every living cell contains within its nucleus small bodies that could be stained with certain colored dyes. They termed these colored bodies "chromosomes." A typical human cell contains within its nucleus 46 chromosomes. During cell division, these chromosomes duplicate (double in number) and, as the cell divides, half of each duplicated chromosome is carried to each daughter cell so that the original chromosome number for that species is retained. Sex cells or "gametes" contain only half the parental number of chromosomes, and when a male and female gamete (egg and sperm) combine, the original number of chromosomes is again restored. It seemed to early investigators that some relationship existed between reproduction and these small colored bodies (chromosomes) within the nucleus. Were these chromosomes, scientists asked, the actual carriers of hereditary traits which were transmitted from parents to offspring? Could they be responsible for dictating the complex biological machinery and chemical reactions that control the synthesis of new cells and dictate such everyday characteristics as blue eyes or brown eyes in offspring? By the 1940s there could no longer be any doubt that hereditary characteristics were carried on the chromosomes within the nucleus of the cell. Furthermore, evidence showed that they carried specific units called *genes* which determined the characteristics of the offspring. But how were genes reproduced in the cell? How did they control heredity and what were they made of? In order to answer these questions, scientists turned to chemical analysis of these nuclear chromosomes.

Chromosomes extracted from the nuclei of cells consist primarily of protein and deoxyribonucleic acid (DNA). Which of these two components carries the genetic information? Oswald G. Avery and others at the Rockefeller Institute extracted purified DNA from the chromosomes of pneumonia bacteria. When they placed this DNA in contact with normally harmless bacteria it caused them to become virulent, indicating that it was DNA and not protein that carries the genetic information. All DNA was found to be composed of the same chemical subunit building blocks called nucleotides, and in 1953 James Watson and Francis Crick, working at the Cavendish Laboratories in Cambridge, England, proposed their now famous determination of the structure of DNA. Finally, in 1967, Arthur Kornberg, starting with basic laboratory chemicals only, chemically synthesized a single strand of DNA that was actually able to duplicate itself in the test tube.

Before proceeding with our question "What is a gene?" let us first examine another question, namely, what are proteins and enzymes? A protein is basically an organic polymer or long chain of small building blocks called amino acids. More than 20 different types of amino acids are linked end to end to form large protein molecules, which are important components of all living organisms. Enzymes, sometimes called organic catalysts, mediate or make possible the thousands of chemical reactions taking place in the living cell. Almost every chemical reaction in a living cell requires a specific enzyme. All enzymes are made of protein, but each specific enzyme differs in the sequence of amino acids making up its protein structure. All the observable characters of living things, such as height or eye color, result from a multitude of specific chemical reactions. These chemical reactions (e.g., those in the production of eye color pigment) are controlled by enzymes, and for a reaction to occur, the proper enzyme must be present. As we shall see later in this chapter, the manufacture of these specific protein enzymes is controlled by DNA in the nucleus of all living cells.

The chromosomes within the nuclei of all cells are composed of long strands of DNA. DNA is made up of the four nucleotide bases—adenine (A), guanine (G), cytosine (C), and thymine (T)—linked end to end (Fig. 5–1). Two of these single DNA strands intertwine to form a coiled double helix pattern, held together by the mutual interaction of weak bonds between the nucleotides. For physicochemical reasons, adenine always pairs with thymine in the opposite strand, and guanine always pairs with cytosine in the opposite strand. The genetic information in DNA consists of three-letter "words" representing the triplet sequence of bases along a DNA strand. Because the four bases can occur in any sequence (such as ACC, AGC, or TCG), there can occur 64 (4×4×4) different "words" or "codons." Another nucleic acid, called messenger ribonucleic acid (m-RNA), composed of adenine (A), guanine (G), cytosine (C), and uracil (U), also occurs within the nucleus of the cell.

During protein synthesis, a segment of the double DNA strand separates, and m-RNA lines up along this strand. Chemical pairing takes place so that adenine of the m-RNA pairs with thymine of the DNA, cytosine pairs with guanine, and uracil pairs with adenine. After a series of three-letter m-RNA molecules line up side by side on the DNA, they link together, separate from the DNA, and move out through the nuclear membrane into the cytoplasm of the cell. Here, the long m-RNA molecule lines up on a cellular protein body known as the ribosome. Meanwhile, a third type of nucleic acid called transfer RNA (t-RNA) picks up amino acids elsewhere in the cytoplasm. These t-RNA molecules also consist of three nucleotide bases, and the sequence of nucleotide bases specifies to which amino acid the t-RNA will attach. After attaching to one of the more than 20 amino acids floating in the cytoplasm of the cell, the t-RNA carrying the amino acid moves to the ribosome, where it aligns with the m-RNA. Again, the nucleotide bases on the t-RNA can align only with their complementary bases on the m-RNA chain. Since A pairs with U, G with C, and U with A, the t-RNA with a base sequence AGU can pair only with the m-RNA sequence UCA. Many t-RNA molecules, each carrying a specific amino acid, line up along the m-RNA chain in the order dictated by the m-RNA base sequence (and originally by

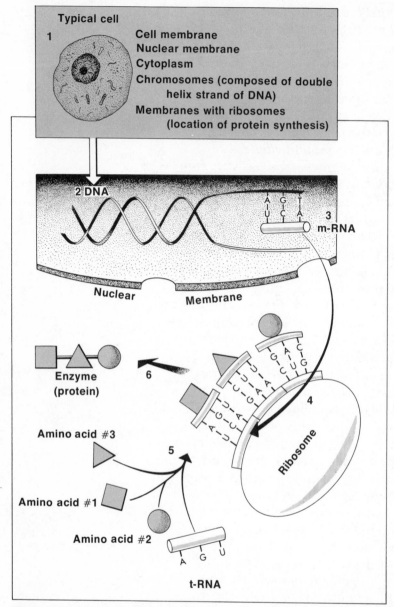

Typical cell

Cell membrane
Nuclear membrane
Cytoplasm
Chromosomes (composed of double
    helix strand of DNA)
Membranes with ribosomes
    (location of protein synthesis)

Figure 5–1 Schematic representation of the role of deoxyribonucleic acid (DNA) in protein (enzyme) synthesis.

(1) Typical cell of a living organism. (2) Two strands of DNA coiled in a double helix. Strands are held together by pairing between nucleotides; adenine (A) pairs with thymine (T) and guanine (G) with cytosine (C). (3) DNA strands separate and messenger ribonucleic acid (m-RNA) aligns on strand with appropriate nucleotide pairing (in RNA, uracil instead of thymine links with adenine). (4) M-RNA moves out on nucleus and into cytoplasm of cell, where it attaches to small protein bodies called ribosomes. (5) Transfer RNA (t-RNA) in the cytoplasm attaches to a specific amino acid and moves to the ribosome, where it pairs with the appropriate nucleotides of m-RNA. (6) Peptide links form between adjacent amino acids to form a "polypeptide" (small protein), which is then released into the cytoplasm. Enzymes (made of protein) formed in this way govern all chemical reactions in living organisms.

the sequence in the nuclear DNA). The amino acids now lying side by side on the ribosome form bonds and link together into short protein polypeptide chains or into larger protein molecules. The protein is then released from the RNA into the cytoplasm of the cell, thus completing the formation or synthesis of a new protein molecule or enzyme.

To return to our original question, therefore, the contemporary view is that *a gene is actually a base (nucleotide) sequence along a certain segment of the DNA molecule that codes for or determines the sequence of amino acids in a newly synthesized protein (or enzyme).* In short, which chemical reactions will occur in an organism depends upon which enzymes are present in quantity. The production of these specific enzymes is controlled by DNA. The gene (the base sequence along the DNA molecule) serves as a source of information, dictating the structure of enzyme proteins.

DNA has the property of self-duplication. During cell division (mitosis) the chromosomes duplicate and separate, and each strand moves to a new "daughter cell." This important sequence of events effectively transfers hereditary information from "mother" to "daughter" cells. In meiosis, the normal (diploid) number of chromosomes is reduced to half (haploid) in four sex cells or "gametes," which carry the genetic information on strands of DNA. In higher animals, including humans, these are the eggs and sperm which, when fertilization takes place, lead to a new combination of genetic material in a growing embryo.

The present large body of organized basic knowledge dealing with the major function of genes opens new horizons, including possibilities for curing cancer or genetic disease and for altering or improving man's genetic make-up. Presented with such powerful new knowledge, man must now face the difficult decision of how he will use it.

BIOENGINEERING—PROMISE OR PERIL?

With newly acquired knowledge of the working of the gene we can now look toward the possibilities of eliminating human genetic defects, controlling our own evolution, and developing a more advanced, more intelligent, and perhaps more human human being. As Isaac Asimov has said, "Knowledge begets knowledge, and the promise of research in molecular biology is fabulous." At present there are more than 1500 known human genetic diseases. Some, such as phenylketonuria, are the result of a single gene deficiency. Phenylketonuria is rare, occurring in only one out of 18,000 live births. Others are the result of multiple factors, and their mode of inheritance is not well known. Diabetes (an inability of the body to manufacture sufficient insulin hormone) is a common example of the latter; it occurs in approximately 4 million people in the United States.

What is the general cause of these inherited genetic diseases? Many result from the cells' inability to synthesize certain required enzymes. Biochemical reactions in living organisms proceed in a stepwise fashion, each step requiring a specific enzyme (Fig. 5–2). Thus, if we start with one chemical compound *(A)* and wish to synthesize another compound *(F)*, a stepwise series of reactions must take place, each mediated by a specific enzyme, and each enzyme is produced by a specific gene (segment of the DNA molecule). Thus, gene 1 may be necessary to produce enzyme 1 and gene 2 to produce enzyme 2. The complete absence of a gene is often lethal. More commonly genes are defective such that one allele* of a gene replaces another allele, leading to a slowdown in the rate of production of an enzyme. For example, in Figure 5–2, if gene 3 on the DNA were defective, enzyme 3 would not be produced in sufficient quantity, and the rate of conversion of compound *C* to compounds *D* and *F* would be too slow. In a complex, delicately balanced system of reactions, even a minor disturbance can have widespread and serious effects. A gene mutation in which enzyme 3 is defective could result in: (1) failure to synthesize required compounds D and F; (2) accumulation of high concentrations of compound *C*; (3) failure to regulate the activities of enzyme 1 by the normal feedback inhibition (see Chap. 3) shown by the arrow from *F* to *A*; or (4) failure to regulate other pathways that may be linked to either *D* or *F*.

How do genetic defects and illnesses manifest themselves in children and adults? A recent review listed 92 human disorders for which a specific enzyme (gene) deficiency has been identified, and it is estimated that more than 25 per cent of children's hospitalization is for genetically related illness.

---

*Genes exist in alternative states called *alleles.* An individual carries only two alleles of any one gene. If these are identical, the individual is *homozygous*; if different, the individual is *heterozygous.* In a population a variety of possible alleles may be found.

GENE

GENE PRODUCT

(ENZYME)

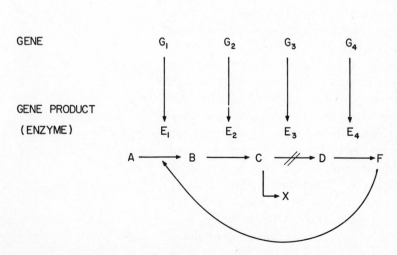

Figure **5–2** A hypothetical pathway for the enzymatic conversion of compound A to final metabolic product F. Compounds B, C, and D are intermediate products. Four different enzymes, E_1, E_2, E_3, and E_4, the products of the corresponding genes G_1, G_2, G_3, and G_4, are required to effect the conversion. The block occurs in the conversion of compound C to D. The concentration of compound F regulates the activity of the first enzyme in the pathway, E_1, in a feedback control loop. (From Friedmann, T., and R. Roblin: Gene therapy for human genetic disease? Science *175*:949–955, 1972. Copyright 1972 by the American Association for the Advancement of Science.)

One such disease, phenylketonuria (mentioned earlier), results from the absence of an enzyme necessary for the metabolism of the amino acid phenylalanine. Since phenylalanine cannot be converted into the next product, toxic accumulations build up within the body and finally lead to convulsions and brain damage. Gaucher's disease produces a deficiency in an enzyme important in the metabolism of brain glycolipids, and results in the death of children one to two years after birth. Hemophilia, the well-known bleeding disorder, results from a deficiency in one of the protein factors necessary for blood clotting. Tay-Sachs disease occurs in children born with the defective form of the enzyme hexosaminidase A. It leads to severe mental retardation and usually causes death by the age of three or four. Down's syndrome (mongolism) results from the presence of an extra chromosome and leads to severe mental retardation. It is the single most common chromosome defect in live-born infants, occurring once in every 600 births.

Should We Eliminate Birth Defects and Genetic Disease?

Effective treatment exists for only a few genetic diseases. In some cases, drugs can block or reduce the accumulation of harmful metabolites. In other cases, genetic blocks resulting in a deficiency of a compound can be treated by administration of the missing compound. Thus, diabetes, which causes a lack of insulin, can be treated by artificial replacement of this insulin; but most genetic diseases do not respond to any known treatment.

If there is a history of genetic disorders in the family, genetic counselors can advise couples on the possibilities that their children will be born with a genetic disease. Ashley Montagu has suggested, "All of us ought to draw up a pedigree of our families, as far back as we can—such pedigrees ought to be as obligatory as a birth certificate or marriage license. A couple contemplating marriage could go over their pedigree charts and from them discover whether they stand in need of genetic counsel."

With amniocentesis, a recently developed technique, doctors can determine the presence of many genetic diseases within the unborn fetus prior to birth. Using this technique, during the early stages of pregnancy a sample of amniotic fluid surrounding the fetus is removed with a hypodermic syringe. The fluid contains cells from the fetus that can then be isolated, grown in tissue culture, and examined under the microscope for chromosomal aberrations, or tested biochemically for the presence or absence of certain enzymes (Fig. 5–3). Dr. Henry Nadler of Northwestern University used amniocentesis to diag-

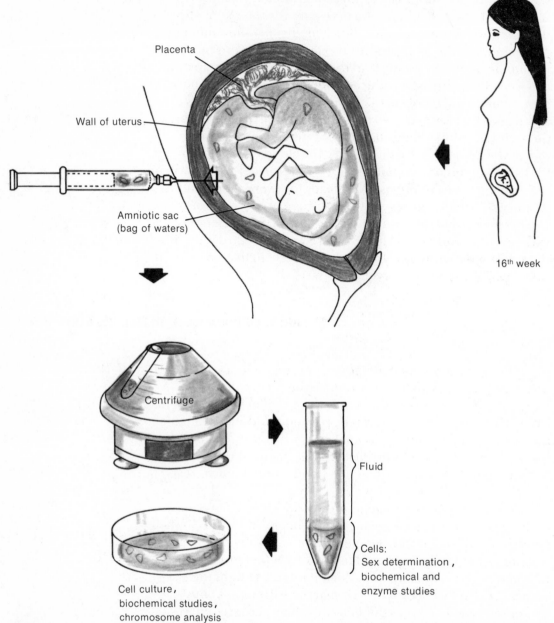

Placenta

Wall of uterus

Amniotic sac
(bag of waters)

16th week

Centrifuge

Fluid

Cells:
Sex determination,
biochemical and
enzyme studies

Cell culture,
biochemical studies,
chromosome analysis

Figure 5–3 The technique of amniocentesis provides a sample of cells and their genetic information from a human fetus around the 16th week of gestation. A sample of fluid (containing fetal cells) is withdrawn from the amniotic cavity and concentrated by centrifugation. Microscopic examination of the cell's chromosomes reveals the sex of the fetus and the presence of some genetic defects such as Down's syndrome. Growth of the cells in tissue culture supplies a large enough quantity for enzymatic studies and further genetic characterization. Care must be taken not to puncture the fetus or placenta.

nose 155 patients believed to have a relatively great probability for Down's syndrome. He found 10 cases of the disease, all of which were confirmed later by examination of the aborted fetuses. John O'Brien and others at the school of medicine of the

University of California, San Diego, used this technique to study the pregnancies of 20 women known to be carriers of fatal Tay-Sachs disease. They found seven affected fetuses, and in all cases these women chose to terminate pregnancy by abortion. Approximately 40 genetic diseases can now be diagnosed prenatally by amniocentesis, and when used by experienced doctors, there appears to be little danger involved in this technique. However, possible side effects have not been studied on a long-term basis. The cost of bearing seriously defective children can be great both in emotional burden placed upon parents and in cost to society at large. For example, 4000 babies with Down's syndrome are born each year in the United States and each requires approximately $250,000 for a lifetime of institutionalized care.

Given that we have the ability at present to eliminate certain genetically defective offspring before birth, the question then arises: is such elimination desirable? Biologists recognize that in a changing environment, genetic diversity is important to the survival of every species—including man. Surprisingly, rather than decreasing the frequency of genetic disease, programs of prenatal detection and abortion might actually bring about the opposite result. Arno Motulsky of the University of Washington describes how this might happen.

Assume that each family has two children, and that it compensates for any defective ones that were aborted by having another. In this case, genes normally lost because of the death or nonreproduction of affected children would be retained in the population by the normal unaffected offspring carriers. This could lead, over many generations, to an increase of 50 per cent in the frequency of carriers of such an undesirable gene in the population. Another question now arises: how do we decide which genetic defects should lead to abortion? Until now, prenatal detection of genetic disease followed by abortion has been used only in cases where serious symptoms or death would arise in the offspring. However, with an increasing number of defects being identified, not all of which are fatal or even serious, we are faced with the question of when to attempt to correct these defects. We are asked to decide, in effect, who shall live. As Theodore Friedmann has said:

> These uncertainties could result in an accentuation of the conflict in our society between personal choice and governmental control, which could possibly come in the form of selected programs of compulsory screening and mandatory abortion for some conditions that are deemed socially intolerable. The obviously dangerous extensions of such a practice would impinge so drastically on our individual liberties as to make them unacceptable and morally unjustifiable.

Artificial insemination and artificial inovulation offer other possibilities for preventing genetic disease. Women whose husbands are sterile or carry defective genes can be artificially inseminated with sperm from anonymous donors (previously checked for genetic defects). In the United States at present, approximately 25,000 women are artificially inseminated each year. In laboratory animals, egg cells have been taken directly from the ovaries, fertilized in a test tube, and reimplanted in the uterus, where they continue to grow. Such *artificial inovulation* could be used for the detection of human genetic defects. By watching the development of a human embryo in the test tube, doctors could detect any genetic deficiencies and decide not to reimplant a defective embryo. In cases where a woman was unable to go through pregnancy, such a test tube embryo could be reimplanted in the uterus of another woman or "donor mother" and carried to term.

In 1961 Daniele Petrucci of the University of Bologna brought *human* eggs and sperm together in a test tube and accomplished fertilization. The resulting embryo continued to grow for 29 days, during which time its heartbeat could be detected, but it soon became quite deformed and the experiment was discontinued. The Chinese newspaper *Jenmin Jin Tao* had the following to say about Petrucci's experiments:

These are achievements of extreme importance, which have opened up bright perspectives for similar research. . . . Nine months of pregnancy is no light or easy burden and such diseases as poisoning due to pregnancy are detrimental to health. If children can be had without being borne, working mothers need not be affected by childbirth. This is happy news for women.

Robert G. Edwards of Cambridge University directs an active research program in the area of human reproductive physiology; and in 1969 he too achieved fertilization of human eggs *in vitro* (in the test tube). Women volunteers, for the most part sterile, came to his laboratory in the hope of gaining the ability to bear children. He injected the volunteers with a hormone (gonadotrophin) causing them to produce many ripe eggs at one time. A small tube was inserted near the ovaries and the eggs were withdrawn by suction, washed, placed in dishes, and after three hours, fertilized with human sperm contributed by the woman's husband. Edwards and his colleagues watched as the sperm penetrated and fertilized the egg and cell division and growth of the embryo began. In 1970, using improved techniques, Edwards's group was able to grow human embryos *in vitro* up to the 16-cell stage (Fig. 5–4). They have hesitated in attempting to reimplant embryos in the uterus until the normality of the embryonic development can be confirmed. Neverthe-

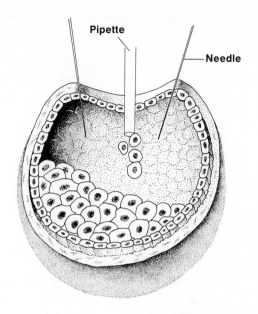

Figure 5-4 Colonization of a mouse embryo with cells from another embryo. Descendants of the injected cells partially colonize the host fetus, resulting in a combination of traits.

less, it is believed that in the near future they will accomplish reimplantation of embryos, leading to the production, after nine months, of a normal human child.

Archie Edwards and Ruth Haler of Cambridge University describe their own researches in human embryology:

> Still further possibilities can be imagined. Eggs fertilized in the laboratory and cultivated to the proteus [early] stage could be transferred back to the mother with an excellent chance of developing normally. Since there would be several from one couple, a degree of selection could be exercised in deciding which one to return to the mother. For example, the sex of the baby could be predetermined.

As these techniques for the elimination of genetic disease become more readily available, will prospective parents become more concerned about the quality of their children? Will they demand the "right" of genetic counseling? Will minor defects now considered to be of little importance become undesirable? Who will decide which defects constitute sufficient reason for abortion and which do not? It was such questions which, in 1971, brought together 80 internationally renowned scholars and scientists to a four-day conference near Washington, D.C. This conference was sponsored jointly by the Institute of Society, Ethics and the Life Sciences and the John E. Fogarty International Center of Advanced Study in Health Sciences (a member of NIH) to discuss "Ethical Issues in Genetic Counselling and the Use of Genetic Knowledge." Some of the questions posed by scientists at this meeting regarding the ethical use of newly acquired genetic knowledge are presented here.

QUESTIONS DISCUSSED AT 1971 CONFERENCE*

Is a genetic counselor's responsibility first to the couple involved, or to society?

What constitutes a "defective" fetus?

When a mother being tested for Down's syndrome in her unborn child turns up negative in that respect, but the test shows the child has an extra Y chromosome—which might predispose him to antisocial behavior as an adult—should the doctor tell her? And if he does, what should she do?

Does an unborn baby have "rights"—including even the right not to be conceived (by cloning) as an identical twin of many others, or of his father? Can a parent ethically give "consent" on behalf of an unborn child—for gene manipulation, for example?

What will be the effect on society, the family, and the individual himself of being able to choose in advance which sex a child will be?

Mass prenatal screening can detect some 130 biochemical abnormalities. Who should do such screening? Which defects should have priority in the search? Should it be voluntary or compulsory?

Who should have access to the information from prenatal tests? How should it be used?

Could a child with genetic defects arising from a rubella infection early in pregnancy sue his mother's obstetrician—or his parents—for "wrongful life" in failing to abort him?

What public policy can and should be worked out to prevent abuses of current and future advances in genetics?

Can Defective Genes Be Corrected?   Within the past five years, scientists have isolated a specific gene from bacteria and completed the chemical synthesis of a single gene in the test tube. These great scientific accomplishments have led to the proposal that "good" DNA, either synthesized chemically or extracted from another organism, could be used to replace "defective" DNA in those suffering from genetic disease. Theodore Friedmann and Richard Roblin of the Salk Institute for Biological Studies have reviewed the prospects for using isolated DNA segments for genetic engineering in human beings, and they believe that genetic modification of human characteristics by incorporation of "exogenous" (foreign) DNA is entirely possible. It has been known for many years that bacteria can transfer traits from one cell to another. Furthermore, when human cells are treated in tissue culture with DNA extracts or certain viruses, they tend to permanently acquire heritable characteristics by physically incorporating the "exogenous" or virus DNA into their own chromosomes. There are problems involved in the incorporation of exogenous DNA into the chromosomes of human cells, but Friedmann and Roblin

*From Peter, W. G. III: Ethical perspectives in the use of genetic knowledge. Bioscience 21(22):1134, 1971.

Virus or foreign
DNA extract

Becomes integrated

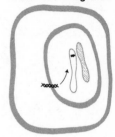

The gene is modified

Figure 5–5 Steps in the genetic modification of a mammalian cell. The added exogenous genetic information may be integrated into the chromosome of the recipient cell and become expressed as a new gene product. (From Friedmann, T., and R. Roblin: Gene therapy for human genetic disease? Science *175*:951, 1972. Copyright 1972 by the American Association for the Advancement of Science.)

believe that these problems can be overcome. For example, rather than using bacterial or viral DNA, some scientists hope to synthesize desirable genes in the test tube and introduce them into genetically defective cells (Fig. 5–5).

In 1973, Nobel prize winner Har Gobind Khorana and his associates at the Massachusetts Institute of Technology synthesized the first completely artificial gene with the potential for functioning inside a living cell. They synthesized two complementary 126-unit polynucleotides into a two-stranded helical gene that codes for the production of a particular bacterial t-RNA (tyrosine transfer RNA). Furthermore, they were able to incorporate the gene into a bacteriophage—a virus that infects bacteria. The virus was then used to introduce the artificial gene into bacterial cells. They found that if mutant bacterial strains lacking the gene to produce the tyrosine transfer RNA were infected with the virus carrying the synthetically produced gene, the mutant strain would assimilate the gene and thus overcome its previously inherited metabolic defect. However, most potentially curable genetic abnormalities in humans involve more complex enzyme deficiencies, and the synthesis of a useful gene that codes for a protein is more complex and difficult to achieve.

With regard to genetic engineering, one could think of many human traits that might be desirable. Rockefeller University biochemist R. D. Hotchkiss has suggested (partially in jest) that the world's food supply problem could be solved by incorporating into the human metabolism an enzyme with the ability to digest cellulose. Humans with this gene would have the ability to digest wood pulp, leaves, and so forth. The real problem, of course, is not simply a shortage of food, but an abundance of people (see Chap. 8). Computers are now being used for determining the structure of chromosomes (see Chap. 3). It has been suggested that in the next century it may be possible to code for the duplication of a specified genetic make-up, have it sent over long distances to another planet, and reconstructed there as a duplicated human being.

Owing to its irreversible and heritable nature, Friedmann and Roblin advise extreme caution in applying the technology of gene therapy. Furthermore, they propose the establishment of certain criteria for the implementation of genetic engineering, including: (1) adequate biochemical characterization of the prospective patient's genetic disorder; (2) prior experience with untreated cases of the same genetic defect, so that the possibility of alternative therapies could be considered; (3) adequate characterization of the exogenous DNA used for replacement; (4) extensive studies on experimental animals; (5) in the case of certain genetic diseases, treatments could first be tested on tissue cultures from the patient without risk to him; (6) the patient, after learning the nature of the proposed treatment and knowing all the risks involved, should give his informed consent to the physician.

Friedmann and Roblin believe that research in the development of gene therapy should continue, but for the foreseeable future, they oppose any attempts at gene therapy applied to human patients because of our present lack of understanding of some of the basic processes of gene regulation. Furthermore, they propose that a sustained effort be made to formulate a "complete code of ethico-scientific criteria to guide the development and clinical application of gene therapy techniques."

Test Tube Babies

Because we now have the technology to fertilize human eggs *in vitro,* will it soon be possible to bypass the natural process entirely and grow human fetuses from conception to term in an artificial environment? Incubators, of course, have been used for years to save premature babies by substituting for some of the functions of the womb and placenta during the last one, two, or sometimes even three months of development. Drs. Warren Zapol and Theodor Kolobow at the US National Heart Institute are now developing an *artificial womb*. Experimenting on lambs, they separate the fetus from its mother during the later stages of gestation, place it in a solution resembling amniotic fluid, and attach the umbilical cord to a series of machines including a pump, a bag of sheep blood, and silicone membranes which remove waste products and supply nutrition. They have been able to keep fetuses alive in this way for several days. Such work indicates that the development of a working artificial placenta and womb for human fetuses may be only a matter of time. Development of such techniques could reduce the number of birth defects by early embryological screening, and allow women who might otherwise have trouble in childbirth to avoid carrying a child altogether. Speculations have even arisen that this future trend might make motherhood and natural childbearing obsolete. What does the public think about these new techniques? A recent Harris poll in the United States entitled "New Methods of Reproduction" indicates that 32 per cent of men and 39 per cent of women would approve of artificial inovulation (reimplanting test tube embryos in the womb for continued development). Considering our society's drive toward convenience in all things, the future possibility of human babies produced entirely in the laboratory appears quite real. It is not hard to imagine the far-reaching implications this might have for the general structure of society (Fig. 5–6).

Should We "Xerox" Humans?   Eggs of certain animals can be made to divide and grow into an embryo without sperm fertilization. Such parthenogenesis (asexual reproduction) can be stimulated by various shock treatments such as the application of extremely cold dry ice to unfertilized eggs. Because eggs contain the female X chromosome, this technique produces only female offspring. If this technique were to be successfully applied to human eggs, multiple female genetic twins would be produced from such eggs without the necessity of fertilization by male sperm.

Another technique of asexual reproduction is called *cloning*. More than 10 years ago John Gurdon destroyed the nuclei of frog eggs by radiation and replaced them with nuclei from the intestinal cell of a tadpole. The eggs, having their full set of chromosomes restored, began to divide and produced a full-grown tadpole that was a genetic twin of the tadpole supplying the nucleus. When and if this technique is applied to human

GOOD MORNING DADDY!

Figure 5-6 Courtesy of Daily Mail, London.

reproduction, we may find a future world filled with identical twins produced by cloning in laboratory test tubes and carried through gestation by donor mothers or hatched in artificial wombs. Will we then deem it necessary as in Aldous Huxley's fictional *Brave New World* (written in 1932) to produce races of alphas, betas, and gammas (workers, pleasure-seekers, and intellectuals) to carry out society's tasks?

Not everyone is greeting the potential arrival of *Brave New World* with open arms. Nobel prize winner James Watson has warned against the dangers of such genetic manipulation. He predicts that if trends proceed in the current direction, a human being born of clonal reproduction may appear within the next 20 to 50 years or even sooner if we actively wish to promote research in this area. Our most sensible course of action, he says, would be to "deemphasize all those forms of research which would circumvent the normal sexual reproductive process." Scientific research is carried on internationally, but limitations on such research could be made by each responsible nation.

If these genetic techniques become routine, should we then actively pursue the development of programs for improving the human race? The problems involved in such "eugenic" programs have been discussed by Ashley Montagu and others. In breeding people of superior intelligence, for example, we may neglect the development of other qualities such as adaptability, integrity, compassion, and a balanced temperament. Today we face the immediate problems of overpopulation, pollution, and war, and problems of new methods of human reproduction appear remote, but as Watson has said, "If we do not think about it now, the possibilities of having a free choice will one day suddenly be gone."

Should Man Achieve Immortality?

As of 1971, thirteen human cadavers were being maintained in the frozen state at −320° C in tanks of liquid nitrogen. These people believed that someday science would find a way to restore them to life and cure the disease that killed them, or that a way would be found (such as cloning discussed above) to produce an identical body from the frozen stored cells of their original body and then implant in the new body the original personality and memory.

Cryobiology (Greek *kryos* = cold) is a new field of scientific study. It deals with the effects of freezing and low temperatures on living organisms. Although there is no evidence that human bodies can be frozen and revived (in fact there is considerable evidence that freezing damages the body), proponents believe there exists at least some possibility that their dream of immortality will come true. Biologists have successfully frozen and thawed human blood and sperm cells as well as mouse embryos. Also, cells of species in danger of extinction (evolutionarily) are being frozen in the hope of reviving them at a later date. Under present laws a human body can be frozen only after it is legally declared dead, but even after a body is "dead," many cells remain alive for hours. Thus, if the body is frozen immediately to very low temperatures, some of the cells may be preserved, but even proponents of freezing human bodies admit that such freezing causes cell damage.

J. K. Sherman of the University of Arkansas School of Medicine believes that attempts being made to preserve human bodies by freezing are "premature, unscientific, and fruitless." We simply do not have enough basic knowledge of the effects of freezing on tissue. Early embryonic stages of mice have been successfully frozen and thawed; but no adult stage of a mammal or other vertebrate has ever survived freezing and thawing. The sincere members of the cryonics societies believe that people have nothing to lose by being frozen after they are dead. But Dr. Sherman believes that we must consider the great expense, false hopes, and disregard for scientific direction that accompany such attempts.

THE EVOLVING CYBORG

With increased miniaturization and sophistication of computers and artificial organs (see Chap. 3), some authors have speculated on the emergence of cyborgs (cybernetic organisms) consisting of a fusion of a human organism or its parts with nonliving artificial devices or machines. Medical science's continuing efforts to prolong human life now increasingly include successful organ transplantation and the use of artificial organs.

The development of the semiartificial man is now rapidly underway. Modern medicine, using new synthetic materials, has developed workable replacements for many body parts (Fig. 5–7). Limited external replacement of damaged portions

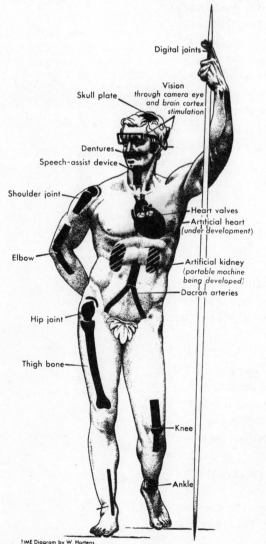

TIME Diagram by W. Hortens

Figure 5–7 The Semi-Artificial Man. (Time, March 18, 1974, p. 56. Reprinted by permission from W. Hortens, Time Magazine, 1974, © Time Inc.)

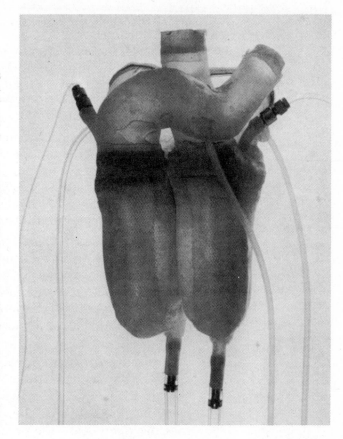

Figure 5–8 Artificial heart designed to fit into a calf's chest is made of Silastic, a smooth silicone rubber. Its structure is closely analogous to that of a real heart. Deoxygenated blood from body enters heart through the vena cava and fills the right atrium (*chamber at top left of photograph, partly concealed by twisting tube*). It is passed on to the right ventricle (*bottom left*). Right ventricle pumps blood to the lungs via the pulmonary artery (*twisting tube at top*). In the lungs, blood is oxygenated. It returns to the left atrium (*top of chamber at right*), then to the left ventricle (*bottom right*). Finally the left ventricle pumps blood into the aorta (*straight tube at top*) and thereby through the circulatory system. The two tubes with metal connectors at bottom are for input of air needed to squeeze the flexible sacs inside the rigid chambers. Other tubes are for gauging pressure and release of residual air at the time the heart is inserted in the body. Dacron mesh forms a framework for the chambers. It is covered by smooth Silastic. (From Kolff, W. J.: An artificial heart inside the body. Sci. Am. *213*(5):38–46, 1965. Copyright © 1965 by Scientific American, Inc. All rights reserved. Photo courtesy of Paul Weller.)

of the body with artificial parts has been possible for centuries. Peglegs have been used since 600 BC; metal hands were made in the sixteenth century; and false teeth were common in the eighteenth century. But unlike earlier externally worn devices, many modern devices are true replacements, permanently implanted within the body. A damaged or diseased thighbone may now be replaced surgically with one made of titanium and other modern alloys; cords of woven dacron in some cases are used to replace damaged tendons; and damaged or arthritic joints may be replaced with joints of plastic and steel. For the circulatory system, tubes of knitted dacron are used to replace blocked or damaged blood vessels, and synthetic heart valves are commonly installed. Researchers are now exploring the development of a completely artificial heart (Fig. 5–8). Patients who have had to undergo laryngectomy (removal of the larynx and vocal cords) may look forward to artificial speech-assist devices now being developed. A small instrument worn outside the upper chest contains a valve system to regulate airflow from one opening in the side of the neck to another in the windpipe, so that expired air activates tissues in the esophagus to produce near-normal speech. Miniaturized artificial kidneys, under development, can be carried in a shoulder case the size of a handbag.

Even blindness may now show some possibility for partial correction. Scientists are developing an artificial vision unit consisting of an implanted subminiature television camera within a glass eye, a tiny computer within an eyeglass frame, and electrodes in the brain to stimulate the visual cortex. Preliminary tests on a few patients indicate that blind subjects assisted by this device have the ability to recognize simple patterns which appear as an outline of light dots.

The US Defense Department is now pursuing research to develop a direct link between the computer and the human mind. They have demonstrated that a man wired to a computer by electrodes placed on his scalp can tell a computer a limited number (seven) of simple commands by simply thinking them. The computer responds to a sophisticated interpretation of electroencephalograph signals (see Chap. 4). Success thus far has been limited. The computer correctly responds to "thought commands" only about two-thirds of the time.

Present day spacemen are closely linked to their mechanical support systems and depend upon them for their very breath of life. Future technologists may include the construction of robots controlled by scientists from earth laboratories, to carry out explorations in the vacuum of outer space or the high pressure deep below the oceans.

In his science fiction book, *The Last and First Men*, Olaf Stapleton describes laboratories where human embryos are treated during development so that the body and lower organs are inhibited in growth while the cerebral hemispheres (centers for thought) are stimulated. After many years of research, a forty-foot superbrain is produced. After copying itself 10,000 times, the many superbrains link together by a radio, take over the world, and use humans as slaves to tend them. Other authors such as J. D. Bernal have even suggested that natural organic evolution of man is now at a dead end. The next evolutionary step may be the emergence of mechanized man—a sharp break with past human evolution. In the not-too-distant future we may find a race of beings many of whom contain artificial parts. Also in the population we may find partial robots containing organs transplanted from human beings. Such cybernetic organisms, or cyborgs, combining both organic and artificially constructed parts, may become commonplace. As scientist-author Isaac Asimov has mused, "Eventually it may not be terribly important to be able to tell the difference between these cyborgs and 'real' human beings."

S. E. Luria of the Massachusetts Institute of Technology foresees four types of application of our new biological knowledge: medical, bioindustrial, social, and military. In medicine we will find increasing prospects for replacing defective genes for the treatment of genetic disease. In the bioindustrial field it may become possible to use direct genetic manipulation instead of selective breeding to produce healthier and more desirable varieties of plants and animals. Socially, pressures could develop to select certain types of inheritable traits. Militarily, we may witness the development of viruses that could affect an enemy population, making it sensitive to poisons, tumors, or even transmissible genetic defects (see Chap. 6).

Aldous Huxley in 1932 portrayed in *Brave New World* a nightmarish utopia based on eugenics, artificial fertilization, twinning induced in a test tube, chemical conditioning of growing test tube embryos, and psychological conditioning of growing children. Many of these techniques (see earlier in this chapter, and Chap. 4) are rapidly approaching reality. Society will have to decide how to use this new knowledge. Who will make the decisions concerning what is desirable or undesirable, superior or inferior? We will be faced, as Luria says, "with the terrifying responsibility of deciding what we—the human race—intend to become."

REFERENCES

Asimov, I. The Genetic Code. New York, The New American Library (Signet), 1962 (paperback). (Assumes no previous chemical knowledge and clearly explains the structure and function of DNA.)

Augenstein, L.: Come, Let Us Play God. New York, Harper & Row, 1969. (Medical ethics.)

Crick, F. H. C.: The genetic code: III. Sci. Am. *215*(4):55–62, 1966. (Describes evidence confirming that the 3-letter language in DNA controls protein synthesis.)

Edwards, R. G., and R. E. Fowler: Human embryos in the laboratory. Sci. Am. *223*:44–54, December 1970. (Description of fertilization of human eggs and culturing of embryos in the laboratory.)

†Freedman, N.: Joshua Son of None. New York, Dell Publishing Co., 1973 (paperback). (Fictional account of cloning.)

*Friedmann, T.: Prenatal diagnosis of genetic disease. Sci. Am. *225*(5):34–42, 1971. (Well illustrated discussion of types of inherited genetic disease, new methods for their detection, and ethical questions involved.)

Friedmann, T., and R. Roblin: Gene therapy for human genetic disease? Science *175*:949–955, 1972. (Prospects for using isolated DNA or viruses for correction of genetic disease; ethical questions involved.)

Grossman, E.: The obsolescent mother. Atlantic Monthly, May 1970, pp. 39–50. (Historical development and present state of technology for artificial inovulation, artificial wombs, and "test tube" babies.)

Husingh, D.: Should man control his genetic future? Zygon *4*:188–199, 1969. (Ethical issues of genetic manipulation in man, problems, and proposed guidelines.)

Huxley, A.: Brave New World. New York, Harper & Row (Perennial Classic), 1932. (The classic portrayal of a frightening eugenic utopia.)

Kolff, W. J.: An artificial heart inside the body. Sci. Am. *213*(5):38–46, 1965. (Manmade hearts have already been placed in animals and the prospect of further progress is good.)

Lederberg, J.: Genetic engineering and amelioration of genetic defect. Bioscience *20*:1307–1310, 1970.

*Lerner, M. I.: Heredity, Evolution, and Society. San Francisco, W. H. Freeman and Co., 1968. (Excellent text on genetics, human evolution and diversity, and social implications of genetic engineering.)

*Recommended further reading
†Fiction

Luria, S. E.: Modern biology: a terrifying power. Nation *209*:406–409, October 20, 1969. (Brief look at expected developments in genetic engineering and the responsibilities they present for scientists and the public.)

*Man into superman: the promise and peril of the new genetics. Time, April 19, 1971, pp. 33–52. (Excellent nontechnical review of biological and genetic engineering and possible effects on society.)

Petchesky, R. P.: Issues in Biological Engineering: A Review of Recent Literature. New York, Columbia University Institute for the Study of Science in Human Affairs, 1969.

Peter, W. G., III: Ethical perspectives in the use of genetic knowledge. Bioscience *21*(22):1133–1137, 1971. (Review of discussions held and questions posed at 1971 meeting of scientists and others.)

Rosenfeld, A.: The Second Genesis; the Coming Control of Life. Englewood Cliffs, New Jersey, Prentice-Hall, Inc., 1969.

Shils, E., et al.: Life or Death: Ethics and Options. Seattle, University of Washington Press, 1968. (Collected essays on medical ethics; see especially P. B. Medawar: Genetic options: an examination of current fallacies.)

Srb, A. M., R. D. Owen, and R. S. Edgar, eds.: Facets of Genetics: Readings from Scientific American. San Francisco, W. H. Freeman and Co., 1970. (Selected well illustrated readings from Scientific American.)

Taylor, G. R.: The Biological Time Bomb. New York, The New American Library (Signet), 1968 (paperback). (Bio- and genetic engineering and their implications.)

The modern men of parts. Time, March 18, 1974, pp. 56–58. (Brief description of new developments in artificial organs.)

Vaux, K., ed.: Who Shall Live? Philadelphia, Fortress Press, 1970. (Collected essays on genetic engineering, abortion and the law, and cloning of humans.)

Watson, J. D.: Moving toward the clonal man: is this what we want? Atlantic Monthly, May 1971, pp. 50–53. (Most sensible course is to deemphasize all research toward circumventing the normal sexual reproductive process.)

Wiley, J. P., Jr., and J. K. Sherman: Immortality and the freezing of human bodies: the case for and the case against. Natural History *80*:12–15, December 1971. (Thirteen people have been frozen in liquid nitrogen in the hope that someday they can be revived.)

*Recommended further reading

PROBLEMS

1. Is genetic counseling very popular? Is it expensive?

2. Will it be possible for science to produce "tailor-made" human beings according to ordered plans?

3. Because of the immune reaction, wouldn't cells tend to reject introduced "exogenous" genes?

4. Is it possible to see a gene or DNA under any microscope? Can DNA actually be seen by x-ray crystallography?

5. What is the definition of death?

6. Are scientists trying to *synthesize* life (i.e., create even the sperm and egg cells), or is this field limited to working with human sperm and egg cells in the test tube?

7. What is meant by the term "inherited genetic disease"?

8. Future programs of "eugenics" have been discussed. What are some dangers involved in controlled "scientific" breeding programs for reducing human genetic disease and physically improving human offspring?

BRAINSTORM

It is the year 2001 and you have been appointed to head the president's advisory council on biological engineering. You are responsible for formulating goals and objectives for the many government laboratories now actively engaged in human genetic engineering. Country X has recently been able to produce vastly superior human beings (resistant to disease and high in intelligence) and there is much sentiment among both politicians and the public that we are falling behind country X in the "gene race." In your new position, how will you proceed?

CONSIDER:
(1) Country X may have expansionist ambitions.
(2) Any program with goals to counter country X may impinge on individual reproductive freedom; that is, your small group might have to decide who could reproduce and who could not.
(3) In the long run, what might be the best policy for your country to pursue regarding genetic engineering?
(4) In your prestigious position, you may be able to influence public opinion, by public statements and education programs.

6

AGGRESSION AND THE TECHNOLOGY OF MODERN WARFARE

The need is not really for more brains, the need is now for a gentler, a more tolerant people than those who won for us against the ice, the tiger, and the bear.

Loren Eiseley

ARE AGGRESSION AND VIOLENCE NECESSARY?

We live in a violent world. Is there a biological drive or instinct behind man's inhumanity toward man? Can we eliminate violence? According to Konrad Lorenz, author of *On Aggression*, there can be little question that aggression and violence once served adaptive purposes for mankind. Early man, a relatively harmless omnivorous creature, lacked the powerful killing abilities of other carnivores. Evolutionarily, large carnivores (such as lions and wolves) developed the ability to instantly kill members of their own species, but evolution also wisely provided them with built-in inhibitions to control such aggression and to prevent them from injuring or killing members of their own species.

In man also, natural selection and evolution provided inhibitory mechanisms to prevent killing members of the same species; however, primitive man invented tools and artificial weapons, and suddenly his killing potential far exceeded his social inhibitions. Primitive apemen used their newfound tools to kill, not only game but also members of their own species.

Thus, beside the fires of Peking man (*Sinanthropus pekinensis*) anthropologists find the roasted bones of his brothers. The history of mankind is a history of technologies of increasingly sophisticated and destructive weapons and means of warfare (Figs. 6–1 and 6–2).

Technologies of violence and warfare developed by man have, in effect, increasingly separated him from his enemies. Inhibitions to violence are ineffective at a distance — the enemy has been dehumanized. In fact, violent instincts may play little part in modern warfare. Humans drop bombs from an altitude of 9000 meters onto a land and people they have little feeling about, or fire ICBMs at others 12,000 kilometers away. Such actions may have their roots in economics and politics rather than in instinctual aggressive psychology.

What Is Aggression?

What do we mean by aggression? Definitions of aggression differ. Konrad Lorenz defines aggression as "the fighting instinct in beast and man which is directed against members of the same species." Others take a broader view and define aggression as an entire spectrum of assertive, intrusive, and attacking behaviors. Given this broader definition, natural selection, by placing a high premium on behavior traits such as

Figure 6–1 A siege catapult. The arm of the weapon is wound down till its point is secured by the catch to be seen beneath it. One end of the winding rope is made fast to an iron bar fixed across the framework of the machine and its other end to the roller. There is, however, no indication of how the arm was cast off from the rope when the former was wound down to the catch, nor is there any form of safety check to be seen on the winding roller.

Figure 6–2 Page from an illustrated version of Virgil printed in Strasbourg in 1502. The Greek stories are illustrated with plates showing sixteenth century styles of armor and weapons (Mungeam Collection). (From Wilkinson, F.: Edged Weapons. New York, Doubleday and Co., 1970.)

amassing property and self-assertion, may still be working today toward increasing aggression in man. Violence (a special type of aggression) has been defined as destructive aggression involving physical damage to persons or property.

A changing environment presents animals, including man,

with a situation of stress, and like other animals, man must adapt to this stress or face the possibility of extinction as a species. Environmental change may render behavior that was formerly adaptive unadaptive or maladaptive. In the past, aggression helped man to protect himself against outside predators or a harsh environment; it was fundamental to his survival. But in the current world of massively destructive weapons, aggression and violence in man may have lost what adaptive value they formerly had.

Marshall F. Gilula and David N. Daniels of Stanford University have reviewed the three major theories of aggression, related them to adaptation, and illustrated the current need for nonviolent ways of expressing aggression. The first theory of aggression, the biological instinctual theory, states that man is a naturally aggressive animal. Proponents of this idea point out that violence or interspecies aggression is rare in other animals, usually occurring only under conditions of crowding. In man, on the other hand, aggression is inevitable, and to reduce violence we must provide channels for the nonviolent expression of aggression. Second is the frustration theory, which proposes that aggression results from interference with ongoing, purposeful activity. Aggression results from frustrations, and aims at removing the obstacles to specific goals. Consequently, a reduction in violence must be achieved through the reduction of existing frustrations. Third, the social learning theory places aggression in the category of socialized behavior patterns resulting from child-rearing practices and other forms of social learning. Aggressive behavior is acquired merely by watching and learning and often by imitation. Reduction of violence requires changes in cultural traditions, child rearing, and parental examples. As Gilula and Daniels point out, these three

'Let's wait. Maybe he kills her or something.'

(Drawing by Weber © 1966 the New Yorker Magazine, Inc.)

theories are not necessarily mutually exclusive and, in fact, are interrelated. Examples of violence may include components from man's inherent aggressiveness and his aggressive responses to frustrated plans, as well as behavior patterns learned in a cultural setting. All three may represent adaptations to coping with stressful situations.

Today the mass media bombard us with one scene of violence after another. One 1969 study found that 85 killings were shown on three television networks within a period of 85½ hours. If the social learning theory (that aggression is largely learned) is true, do the mass media teach violence? Mass media–portrayed violence surely has an effect on almost everyone, but particularly on children. The average American between the ages 3 and 16 spends more waking hours watching television than attending school. Sixth graders spend 80 per cent of their viewing time watching "adult" programs, and by the eighth grade, children favor crime programs above others.

After many studies, the theory that watching television violence serves as an outlet or "catharsis" for potential aggressive behavior has been largely discredited. There is considerable evidence that television in its present form teaches violence, but the actual performance of violence by the viewer may depend more upon the environment. Inhabitants of ghetto areas or others with frustrated goals do not have to look far for adaptive ways of channeling their aggression into violence. As Gilula and Daniels say, "In teaching that violence is a good quick way to get things done, television and other media teach that violence is adaptive behavior." Thus, one means of reducing violence might be to reduce the number of mass-media displays of violent behavior.

Society provides certain sanctioned forms of violence, including capital punishment and war. The exercise of these violent means is quite often arbitrary and under the control of powerful political leaders. The causes of war are complex and certainly not well understood. They may involve components of socially learned prejudice and fear, and biologically acquired aggressive instincts.

Violence, not aggression per se, is the primary threat to mankind. New nonviolent channels for the expression of aggression should be found. These might include public ombudsmen to receive citizen complaints or to negotiate conflicts. Means of nonviolent protest and negotiation should be taught to those involved in social change. Violence, although firmly rooted in man's biological and evolutionary heritage, has become, with the development of modern weapons, increasingly maladaptive; our very survival may depend on how well we can channel and redirect these aggressive instincts.

Can man survive in a world filled with weapons of mass destruction of his own creation? The proposition that we all have to die someday does not mean that we all have to die the same day. C. Wright Mills in his book *The Causes of World War Three* suggests several ways in which another world war could be averted. First, scientists and intellectuals should take a lead in declaring their intention to work toward peace; they should denounce secrecy in their research. They should refuse to become servants of military authority, refuse to work on weapons projects, and boycott all research directly or indirectly relevant to the military. Scientists are beginning to assume the responsibility of making facts available to the public on matters of science, technology, and public policy (see Chap. 10). With the creation of modern weapons of mass destruction, man has developed the potential to destroy his entire species. Given this potential, violence can hardly be considered adaptive behavior in modern man. Today men must face the truth that they must either live together or die together.

THE DEVELOPMENT OF ATOMIC WEAPONS

During the early part of the twentieth century, scientists accumulated an organized body of basic knowledge concerning the structure of the atom and the nature of subatomic particles (see Chap. 2). Between the years 1918 and 1923 Lord Rutherford led an active research group at the Cavendish Laboratories in Cambridge, England. By bombarding atoms with subatomic particles, this group contributed much to current understanding of subatomic physics. At the same time, another young group of physicists were studying under A. J. W. Sommerfeld in Munich. This five-year period represented an exciting time for basic theoretical research in nuclear physics. As the American physicist J. Robert Oppenheimer later wrote:

It was a period of patient work in the laboratory, of crucial experiments and daring action, of many false starts and many untenable conjectures. It was a time of earnest correspondence and hurried conferences, of debate, criticism, and brilliant mathematical improvisation. For those who participated it was a time of creation. There was terror as well as exaltation in their new insight.

These were the beautiful years for atomic physics. In the years 1923 to 1932, world research in basic nuclear physics centered at the University of Göttingen in Germany. Scientists, including Oppenheimer (later to be known as the father of the atomic bomb), came from all over the world to study at Göttingen. In fact, most of the Americans who later worked on the atomic bomb project studied here during this period.

They all sensed the far-reaching importance of their research in terms of a basic understanding of the physical world; however, few could imagine the enormous practical importance or applicability later to be found for their research.

Ever since Einstein's formulation of the interchangeability of matter and energy ($E = mc^2$), scientists had suspected that nuclear reactions might have been occurring in our sun. Houtermans and Atkinson constructed a mathematical theory of the thermonuclear reactions occurring in the sun involving the fusion of lightweight hydrogen atoms. Little did they dream of the future sinister possibilities for harnessing this reaction for the production of thermonuclear hydrogen weapons. The fame of the University of Göttingen as a center for nuclear research spread throughout the world; however, a collision between research and politics soon occurred. Many of the physicists at Göttingen were of Jewish background, and with the rise of Nazi Germany and the subsequent persecution of Jews, many left Göttingen and fled to the United States, Great Britain, or elsewhere. In 1933 Albert Einstein came to the United States to work at Princeton University. This single fact drew many more physicists to the United States, and Göttingen ceased to be the world center for atomic research.

The Discovery of the Neutron

A major breakthrough in nuclear physics occurred when James Chadwick at Cambridge University discovered the neutron*—the key to atomic fission (Fig. 6–3). Few scientists rea-

---

*A neutron is an elementary atomic particle that can penetrate the barrier of negatively charged electrons surrounding the nucleus of an atom because the neutron is electrically neutral and is not repelled. It can thus be used to bombard and split the nucleus, causing a tremendous release of energy.

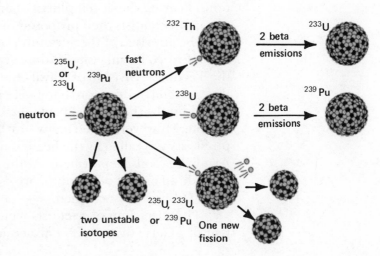

Figure 6–3 A chain reaction. A thermal neutron collides with a fissionable nucleus and the resulting reaction produces three additional neutrons. These neutrons can either convert nonfissionable nuclei, such as ^{232}Th, to fissionable ones or cause additional fission reactions. If enough fissionable nuclei are present, a chain reaction will be sustained. (From Jones, M., et al.: Chemistry, Man and Society. Philadelphia, W. B. Saunders Co., 1972.)

lized the significance of this discovery, but soon they were speculating on the possibility of producing nuclear chain reactions in certain elements by bombardment with neutrons. Paul Langevin, one of the few to realize the serious consequences of this research, when asked if he was worried about Hitler, replied:

I'm much more worried about something else. It is something which, if it gets into the wrong hands, can do the world a good deal more damage than that fool who will sooner or later go to the dogs. It is something which — unlike him — we shall never be able to get rid of: I mean the neutron.

Leo Szilard, a Hungarian physicist, fearing the possibility that this newfound knowledge might fall into the wrong hands, approached a number of atomic research workers in 1935, asking them whether or not they would refrain, at least for the time being, from publishing any future results of their investigations. In a letter to Irène Joliot-Curie in February, 1939, Szilard stated,

Obviously, if more than one neutron were liberated [in the disintegration of uranium], a sort of chain reaction would be possible. In certain circumstances this might then lead to the construction of bombs which would be extremely dangerous in general and particularly in the hands of certain governments.

By "certain governments," Szilard was referring to Nazi Germany and the Axis powers.

By refraining from publishing their findings, Szilard and other physicists hoped to keep the results of their work from the Axis powers. Work progressed, nevertheless, and several researchers published papers dealing with atomic chain reactions. Fearing the development of an atom bomb by Hitler, a group of scientists then proposed that the American government be apprised of the "uranium situation." On March 17, 1939, Enrico Fermi visited Admiral S. C. Hooper, director of the Technical Division of Naval Operations, to inform him of the possibility of an atom bomb. He made little impression on the admiral — the possibility of an atomic bomb seemed too remote. However, in Germany in 1939 a meeting of nuclear physicists was called by the head of the research division of the Army Weapons Department. At the same time, the Germans forbade all export of uranium ore from Czechoslovakia, which they had recently occupied.

The suspicions of scientists seemed confirmed, and the fear of a German atomic bomb accelerated within the scientific community. Fearing that the United States might be caught by surprise by a German bomb, Szilard and other scientists decided that they should take the matter to the president of the United States. They drafted a letter, signed by Einstein, which they delivered to President Roosevelt in October, 1939. Roosevelt decided that the atomic research project advocated by the scientists required action, but skepticism about the practicability of such a project continued, and scientists encountered frustration and difficulty in attempting to procure funds from the government for atomic research.

The Manhattan Project

Finally, on June 6, 1941, just one day before the Japanese attack on Pearl Harbor, the government decision was made to supply substantial support and technical resources for the pursuit of an atomic weapons program. J. Robert Oppenheimer became scientific head of this undertaking, termed the Manhattan Project and General Leslie Groves was appointed army head of the project. The project was carried out in several areas. Basic theoretical research on nuclear chain reactions was pursued at the University of Chicago and at the University of California. At Oakridge, Tennessee, scientists and technicians worked on the physical mechanisms for the construction of the bomb as well as on the production of the fissionable uranium-235 necessary to produce an atomic explosion. A technique for producing fissionable material from another element, plutonium-239, was discovered to be theoretically possible and scientists decided to pursue the development of both a uranium and a plutonium bomb simultaneously. At Hanford, Washington, facilities were constructed for the production of this radioactive plutonium. Finally, a secluded research and testing site was chosen near Los Alamos, New Mexico.

The Manhattan Project grew in size, drawing some of the most prominent physicists of the Allied countries. Eventually, 150,000 persons were employed on the project, but top secrecy was enforced at all levels and only about 12 persons had an overall view of the nature of the project. In 1945, as work on the project continued, Allied troops entered Germany and somewhat surprisingly found little evidence of German research directed toward the construction of an atom bomb.

With Germany practically defeated, and with the knowledge that the Japanese were in no position to develop any atomic weapons, many scientists began to oppose further research toward construction of an atomic bomb by the United States. However, the Army and General Groves pushed for further development of the atomic bomb and ultimately its use on Japan. In the spring of 1945, a group from the Manhattan Project was given the task of selecting a target in Japan. A panel of nuclear physicists at that time was asked not *whether* the bomb should be used but only *how* it should be used.

A group of the Chicago physicists working on the project who were now opposed to its further development, sent a petition to the secretary of war on June 11, 1945, suggesting that the atom bomb first be demonstrated to Japanese observers. Their petition was ignored, final preparations were made, and a test bomb was exploded at Los Alamos, New Mexico. Following the test explosion, opposition among scientists to the use of the bomb grew, and 67 prominent atomic scientists signed a petition urging that the bomb not be used against Japan without offering the Japanese a prior demonstration and an opportunity to surrender. Furthermore, they urged the government to start immediately to secure international agreements to limit and control the new atomic weapon. Their petition never reached the higher levels of the United States government.

Ironically, even before the bomb was used, United States Intelligence had become convinced that Japan would surrender within a few weeks. The State Department was aware of definite Japanese "peace feelers" but was mostly unaware of its own government's intention to use the atomic bomb. On the other hand, scientists and army officers at Los Alamos proceeded with plans for dropping the bomb, unaware of the recent Japanese peace feelers.

The atomic fission bomb that was dropped over Hiroshima in 1945 had an explosive power equivalent to 20,000 tons of TNT. In 1952, the first thermonuclear fusion bomb (hydrogen bomb) was tested at Bikini Atoll in the central Pacific. The explosive power of a present day hydrogen bomb is equivalent to between 10 and 100 million tons (megatons) of TNT — more than 1000 times the power of the bomb dropped on Hiroshima. Theoretically, there is no upper limit to the hydrogen bomb's size and destructive capacity. Bombs of several megaton capacity can be carried thousands of miles by intercontinental ballistic missiles.

Our nuclear age was aptly characterized by General Omar Bradley: "Ours is a world of nuclear giants, and ethical infants."

NUCLEAR ENERGY

Einstein's equation (see also Chap. 2) $E = mc^2$, indicates that a small amount of matter contains a tremendous amount of energy. For example, one gram of matter, if completely converted to energy, would yield 9×10^{20} ergs (357 million calories) or enough energy to heat the average home for 1000 years. *Nuclear fission* (splitting of the atom) occurs when a thermal neutron collides with certain "fissionable" heavy nuclei of uranium or plutonium (^{235}U, ^{233}U, or ^{239}Pu). The splitting of the nucleus produces two smaller nuclei, two more neutrons, and a large quantity of energy. The neutrons released can produce more fission and release more neutrons, resulting in a chain reaction (Fig. 6–3).

A chain reaction occurs when a certain quantity (called the *critical mass*) of fissionable matter is brought together. In the atomic bomb several subcritical masses are kept separate until detonation, when they are driven together. The explosion results in the release of radioactive fragments over wide areas and tremendous amounts of energy in the form of heat as high as 10 million°C.

Nuclear fusion occurs when light nuclei such as hydrogen (H), helium (He), and lithium (Li) are combined or fused, resulting in a decrease in mass and tremendous release of energy. For example, tritium can be fused with deuterium to form helium, a neutron, and energy.

$$(1) \quad {}^3_1H + {}^2_1H \rightarrow {}^4_2He + {}^1_0n + 17.6 \text{ mev}^*$$

Fusion reactions, however, can only be achieved at temperatures of 100 million °C or more.

The high temperatures required for the detonation of a hydrogen bomb are achieved by exploding a fission (atomic) bomb first to "trigger" the hydrogen bomb. In one type of hydrogen bomb, for example, a jacket of lithium deuteride (^{6}Li^{2}H) is placed around an ordinary ^{235}U or ^{239}Pu fission bomb. The energy from the fission bomb splits the lithium into tritium and helium.

$$(2) \quad {}^6_3Li{}^2_1H + {}^1_0n \longrightarrow {}^3_1H + {}^4_2He + {}^2_1H$$

The temperature produced by this fission is high enough to bring about the fusion of tritium and deuterium in reaction (1) above. A 20-megaton bomb contains about 136 kilograms of lithium deuteride as well as considerable amounts of plutonium and uranium.

Peaceful uses of controlled fission and fusion are discussed in Chapter 7.

*Million electron volts.

THE DEVELOPMENT OF CHEMICAL AND BIOLOGICAL WEAPONS

Chemical warfare is not new. The Greeks, for example, during the Peloponnesian war (431–404 BC), burned large masses of pitch and sulphur in order to smoke out the enemy during sieges. The poisoning of water wells, long used as a technique of warfare, was prohibited by the Romans and declared unethical. With rapid increases in industrial technology during the nineteenth century, many countries became concerned about the possibility of chemical warfare. At the second Hague Peace Conference, in 1907, the use of poisonous gases in warfare was explicitly forbidden. Nevertheless, only a few years later new forms of chemical warfare were initiated. During World War I, the French attacked the Germans with tear gas, claiming it had not been outlawed by the Hague Peace Conference. The Germans, believing that it had, retaliated in 1915 with chlorine gas, producing 15,000 casualties, including 5000 dead. The Allies then escalated with phosgene gas, a substance six times more deadly than chlorine. Estimates indicate that at least 125,000 tons of toxic chemicals were used during World War I, and that these produced approximately 1.3 million casualties and 100,000 dead. After World War I, the United States took the lead in promoting the Geneva Protocol of 1925, which banned "the use in war of asphyxiating, poisonous, or other gases" and prohibited bacteriological warfare. However, under opposition from certain politicians, from the American Legion, and from the American Chemical Society (which more recently reversed its position), the United States did not ratify the Geneva Protocol at that time.

Great advances in basic knowledge of genetics and microbiology, coupled with applied chemical and biological warfare (CBW) research, have led to contemporary biological and chemical weapons far more toxic than those used in World War I (Tables 6–1 and 6–2). Prior to 1969, for example, the United

TABLE 6–1   Biological Warfare Agents*

Agent	Diseases	Incubation Period (days)	Effect of Treatment	Contagiousness
Viruses	Eastern equine encephalitis	5 to 15	None	By vector
	Tick-borne encephalitis	7 to 14	None	By vector
	Yellow fever	3 to 6	None	By vector
Rickettsiae	Rocky mountain spotted fever	3 to 10	Good	By vector
	Epidemic typhus	6 to 15	Good	By vector
Bacteria	Anthrax	1 to 5	Moderate	Low
	Cholera	1 to 5	Good	High
	Pneumonic plague	2 to 5	Moderate	High
	Tularemia	1 to 10	Good	Low
	Typhoid	7 to 21	Good	High
Viruses	Chikungunya fever	2 to 6	None	By vector
	Dengue fever	5 to 8	None	By vector
	Venezuelan equine encephalitis	2 to 5	None	By vector
Rickettsiae	Q fever	10 to 21	Good	Low
Bacteria	Brucellosis	7 to 21	Moderate	None
Fungi	Coccidioidomycosis	7 to 21	Poor	None

*Biological agents are shown by category and effect; those that could be expected to cause death are at top, and agents that might be used to cause incapacitation are at bottom. Contagion "by vector" means transmission by certain species of mosquitoes or other insects. (From Meselson, 1970.)

TABLE 6–2 Chemical Warfare Agents

Common Name	Chemical Name	Formula	Properties and Symptoms	Description
Chlorine gas	Chlorine	Cl_2	Lethal pulmonary edema and related respiratory afflictions.	Highly poisonous gas used in World War I; protection by gas mask.
Mustard gas	bis-(2-chloroethyl-) sulfide	CH_2CH_2Cl — S — CH_2CH_2Cl	Blistering and lethal lesions, vesication, and blindness.	Absorbed through the skin. Used in World War I.
Nerve gas (VX) (one type)	Probably is ethyl-S-dimethylamino-ethyl-methylphosphono-thiolate	H_3C, H_5C_2O — P(=O) — S—CH_2CH_2—$N(CH_3)_2$	Blocks acetylcholine esterase enzyme necessary for nerve transmission leading to paralysis of both voluntary and involuntary muscles and death.	Airborne concentration of a few mg/m^3 is almost instantly fatal.
White phosphorus	White phosphorus	P_4	Inflicts severe burns; particles must be surgically removed. Inhaling the vapor leads to decay of the bone structure, particularly around the nose and jaw. A cumulative poison.	Incendiary, delivered by shells and land mines. Agent ignites spontaneously on contact with air. Used in Indochina war.
Napalm B (new type)	50% polystyrene 25% benzene 25% gasoline		Burns form disfiguring and deep-seated scars, which may become cancerous. May also kill by asphyxiation.	Incendiary delivered mostly by aerial bombing. Earlier type used in World War II bombing of Japan. Used in Korean and Indochina wars as well as in the Middle East.
Tear gas (CS) (one type)	o-chlorobenzalmalonitrile	(o-Cl-phenyl) — C=C with C≡N groups	Skin and respiratory irritant. At high concentrations in enclosed environment can cause burns and asphyxiation.	Pungent, peppery odor detected at 0.05 mg/m^3. Used for riot control and by US in Indochina war, 1966 to 1970.
Psychochemicals BZ (one type)	Believed to be the phenyl glycolate ester of 3-quinuclidinol but exact structure is secret.	(phenyl) — C(OH)(R) — C(=O) — O — (ring N) BZ?	Effects unpredictable, but may include loss of time sense, hallucinations, perceptual distortions, etc. Incapacitating agent.	Rumors of US experimental use during Vietnam war. Effective at extremely low concentrations.
Herbicides agent orange	50/50 mixture of n-butyl esters of 2,4-dichloro-phenoxy-acetic acid (2,4-D) and 2,4,5-trichloro-phenoxyacetic acid (2,4,5-T)	Cl—(ring)—O—CH_2 COOH with Cl; Cl, Cl—(ring)—O—CH_2 COOH with Cl	In dicotyledonous plants, interferes with the action of auxin, a general plant hormone; at higher concentrations, causes leaves to fall, and death of plant.	First used in warfare by US in South Vietnam 1961 to 1971. Used in lower concentration as domestic weedkiller.
agent blue	Cacodylic acid	H_3C—As(=O)—CH_3 with OH	Active against monocotyledonous plants such as rice. Also toxic to humans.	Contains arsenic; not approved by USDA for agricultural use. Used by US in South Vietnam "resource denial" program.
agent white (Picloram or Tordon)	4-amino-3,5,6-trichloro-picolinic acid	NH_2, Cl, Cl, Cl—(ring N)—COOH	Interferes with hormone systems in dicotyledonous plants.	Less susceptible to drift when dispensed from air; much more persistent (decades) in soil and environment than agent orange.

States engaged in a comprehensive secret research program in the development of chemical and biological weapons. The Army Chemical corps advocated chemical and biological weapons as being humane alternatives to the destructive violence of nuclear weapons, and pointed out the relative cheapness of their production. By 1964 the United States government was spending $458 million per year on chemical and biological warfare research. Fort Detrick, Maryland, maintained one of the world's largest facilities for conducting research on the effects on animals of pathogenic organisms and chemicals.

Nerve Gas

Some of the most toxic agents developed are the nerve gases, which kill when they are inhaled or when they are deposited as droplets on the skin. They operate by interfering with the transmission of nerve impulses across synapses. One nerve gas, known as VX, when spread as an aerosol spray, settles to the ground and can kill any animal it touches within a period of minutes. Contact with even a small droplet (0.2 mg) of VX can be fatal. During 1968, tests at the United States Army station at Dugway, Utah, 6400 sheep were killed when an aerosol of VX drifted off the testing area. The chance of accidental harm to civilian populations during shipment and storage of nerve gases has stimulated the development of new *binary weapons*. In a binary weapon, two relatively harmless agents remain separate until a missile is actually in flight. Then, by remote signal, they are combined to form an extremely lethal agent.

Incapacitating Agents

In addition to toxic nerve gases, modern chemical weapons include incapacitating agents such as tear gas. The principal short-term incapacitating agent now in military use is called CS. It has been used both in riot control and in the Vietnam war, in grenades, rockets, artillery shells, and cluster bombs (Fig. 6–4).

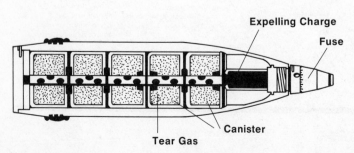

Figure 6–4 Artillery shell designed to deliver CS irritant has the agent in canisters that are expelled when the fuse detonates the shell. The shell portrayed, which is for a 155-millimeter weapon with a range of 15 kilometers, is 60 centimeters long, weighs 44 kilograms and carries 4.4 kilograms of CS. In a normal artillery barrage several weapons are fired, delivering a number of shells to target. (Adapted from Meselson, M. S.: Chemical and biological weapons. Sci. Am. 222(5):15–25, 1970.)

Exposure to CS causes intense pain in the eyes and upper respiratory tract, then progresses to the lungs, where it produces feelings of suffocation and acute anxiety. *Without excessive exposure*, symptoms pass within a few minutes after the exposure ends. However, when applied in high concentrations in enclosed areas, even such an "incapacitating agent" can become lethal. In the Vietnam war, high concentrations of CS were pumped into underground bunkers and shelters, resulting in the death of children who had taken refuge within. The United States used more than 6.3 million kg of this incapacitating gas in Vietnam. Some have argued that such nonlethal incapacitating chemical weapons represent a more humane form of warfare. However, this can hardly be true, since they are generally used in conjunction with conventional weapons (for example, to force the enemy from concealment to open ground where they face the deadly fire of conventional weapons).

Various psychochemicals or hallucinogens have also been suggested as warfare agents that could cause temporary disability by disrupting normal patterns of behavior. Extremely small doses of LSD could produce mental imbalances lasting for periods of 10 hours. The US army has investigated the potentials of a psychochemical known as BZ (Table 6–2). It is difficult to predict what effects an attack with psychochemical agents might have on a large population.

Legality of Chemical and Biological Warfare

Jozef Goldblat has asked whether nonlethal weapons such as tear gas and herbicides are permitted weapons. The United States maintains that the Geneva Protocol of 1925 does not prohibit the use of tear gas in war, and claims that herbicides are not covered by the Protocol. However, Goldblat's analysis of the League of Nations debate during the formation of the Geneva Protocol indicates that international law does indeed prohibit the use of both tear gas and agents affecting plants in warfare. Furthermore, the United Nations has attempted to make clear that the prohibition contained in the Geneva Protocol applies to the use in war of all chemical, bacteriological, and biological agents (including tear gas, herbicides, and other harassing agents). Beyond such questions of legality, the primary danger inherent in the use of *any* biological or chemical agent (even tear gas) in warfare may lie in the possibilities for prompting escalation to more deadly agents.

Much of United States biological warfare research has consisted of attempts to breed pathogenic organisms with desired characteristics and to produce an immunity for one's own population and troops. Much of this research became part of the secret literature maintained by the Department of Defense. What are the advantages of CBW as seen by its proponents? First, it is much cheaper to develop than other types of modern weapons, and second, unlike totally destructive nuclear weapons, it destroys life without destroying property.

In November, 1969, President Nixon announced major cutbacks in research on chemical and biological weapons. He said that the United States would never be the first to use lethal or incapacitating chemical weapons and would not use biological weapons under any circumstances, even in retaliation. Under the new policy, the United States would destroy existing stocks of germ and toxin weapons, and would no longer engage in their development, production, or stockpiling. These latter stipulations, however, did not pertain to chemical weapons, only biological weapons.

Effects of Chemical and Biological Weapons

In 1969, a United Nations study group made a special report on chemical and bacteriological (biological) weapons and the effects of their possible use. As the report points out, many differences exist between various chemical and biological warfare agents. These differences include (1) potential toxicity, (2) speed of action, (3) duration of effect, (4) specificity, (5) controllability, and (6) residual effects. Chemical warfare agents, for example, are generally less toxic on a weight basis than biological agents, since a very small amount of the latter can multiply and spread over a large area, causing many casualties. With regard to their speed of action, some agents, such as certain nerve gases, may act within seconds, whereas some bacteriological agents may require days of incubation before they take effect. Biological agents may have a high degree of specificity for the host they infect; chemical agents are less specific (Table 6–3).

Biological warfare agents that have been developed and studied include various bacteria, viruses, and fungi (Table 6–1). In most cases, these agents are designed to be spread from the air by aerosol spray. Troops on the ground might obtain some protection from these diseases by gas masks and clothing, but in some cases persistent spores might contaminate the area for weeks or months. United Nations reports estimate that using an aerosol spray under certain conditions could contaminate an area as large as 100,000 square kilometers.

Anthrax, an animal disease that rarely infects man, is an example of one biological warfare agent. The pulmonary (lung) form is quite severe, and if not treated early, will cause death within two or three days. The anthrax bacteria form

TABLE 6–3 Comparative Estimates of Disabling Effects of Hypothetical Attacks on Totally Unprotected Populations Using a Nuclear, Chemical or Bacteriological (Biological) Weapon That Could Be Carried By a Single Strategic Bomber*

Criterion for estimate	Type of Weapon		
	Nuclear (one megaton)	Chemical (15 tons of nerve agent)	Bacteriological (biological)** (10 tons)
Area affected	Up to 800 km²	Up to 60 km³	Up to 100,000 km²
Time delay before onset of effect	Seconds	Minutes	Days
Damage to structures	Destruction over an area of 100 km²	None	None
Other effects	Radioactive contamination in an area of 2500 km² for 3–6 months	Contamination by persistence of agent from a few days to weeks	Possible epidemic or establishment of new endemic foci of disease
Possibility of later normal use of affected area after attack	3–6 months after attack	Limited during period of contamination	After end of incubation period or subsidence of epidemic
Maximum effect on man	90 per cent deaths	50 per cent deaths	100 per cent morbidity: 25 per cent deaths if no medical intervention
Multiyear investment in substantial research and development production capability†	$5–10 billion	$1–5 billion	$1–5 billion

*From United Nations Report No. E. 691. 24. 1970.

**It is assumed that mortality from the disease caused by the agent would be 50 per cent if no medical treatment were available.

†It is assumed that indicated cumulative investments in research and development and production plants have been made to achieve a substantial independent capability. Individual weapons could be fabricated without making this total investment.

resistant spores which live for many years, contaminating areas and constituting a persistent danger. Large quantities of this bacteria can easily be grown in a laboratory and dispensed in aerosol sprays over heavily populated areas where the soil would remain contaminated for a long period of time.

Plague, under natural conditions, is carried by small rodents; the microbe is transmitted by fleas to man. When inhaled by man, these microbes produce pneumonic plague after a three- to five-day incubation period, resulting in severe general symptoms, and, if untreated, death occurs within two to three days. It is extremely contagious, even by contact. Again, large quantities of these bacteria can be grown, freeze-dried, and spread by aerosol spray over large areas of population. No effective vaccine is known against this type of disease.

Q fever, another natural disease of animals (occasionally passed to man from sheep, goats, and cattle), is also a biological warfare agent. Its potential use in biological warfare has been studied on human volunteers, who were found to be highly susceptible. Symptoms include a severe attack of an influenza-like illness, followed by high fever and joint and muscle pains, which may be followed within five or six days by pneumonia. As the United Nations report states, "A Q-fever aerosol could produce an incapacitating effect in a large proportion of the population of an attacked area, the infective agent could persist in the environment for months and infect animals, possibly creating natural foci of infection."

Venezuelan equine encephalitis virus (VEE) is transmitted in nature by mosquitoes from animals to man. The disease is generally not fatal, but concentrated aerosols could be expected to incapacitate a very high percentage of the population exposed. Finally, aerosols of tularemia and yellow fever have also been considered as possible biological warfare agents.

Chemical and biological agents have been developed that can specifically destroy crops and other plants. Widespread outbreaks of devastating plant diseases that occur naturally have led to the suggestion that plant pathogens be used for military purposes. These pathogens include rust fungi that attack cereal crops and potato beetles that infest potato-growing areas. Large-scale destruction of crops by such organisms would deprive an enemy (as well as the civilian population) of available food resources. Destruction of crops by either biological or chemical agents, however, produces its most severe effects on civilians. When food is short, soldiers get whatever food there is and helpless people and children are the first to suffer.

The United Nations report on CBW lists 22 different viruses, bacteria, and fungal agents that have been considered as possibilities in biological warfare. The report concluded that the existence of chemical and bacteriological (biological) weapons contributes to international tension and that their further development spurs the arms race without contributing to the security of any nation.

Frightening future technologies of "genetic weapons" are now being considered. These include, for example, ethnically specific chemical or biological agents designed to incapacitate people of specific race groups. The theory is based on race-specific differences in genetic make-up and enzyme constitution.*

ECOLOGICAL CONSEQUENCES OF MODERN WARFARE

Modern weapons provide man with the capacity to destroy not only himself but also to destroy the earth. Gordon Orians and E. W. Pfeiffer made a firsthand assessment of the ecological effects of the war in Vietnam including the effects of defoliation, bombing, and other military activities. This particular war, more than any other, had major destructive effects on the environment. These two researchers gathered information from personal observations and interviews with military personnel, government agencies, and plantation owners.

Herbicides

Herbicides (defoliants) are agricultural chemicals which in high concentrations poison plants and cause them to lose their leaves or die. Little is known at present concerning their long-term effects on animals or on the general ecology of an area. They are used in warfare to increase fatalities by denying food to the enemy, and to deprive him of cover and concealment provided by natural forest growth. The United States has argued that the use of herbicides is not prohibited by the Geneva Protocol of 1925; and they were used over large areas of Vietnam by the United States beginning in 1961. A peak was reached in 1967 when about 600,000 hectares were defoliated, and estimates indicate that 20 to 25 per cent of the forests of Vietnam were sprayed more than once. There are three princiapl anti-plant warfare agents, designated orange, white, and blue (Table 6–2). The concentrations of defoliants used in Vietnam were much greater than those intended for agricultural weed-control and other civilian operations. Mangrove forests are extremely susceptible to defoliants, and one application of agent orange was sufficient to kill most of the trees. Estimates indicate that trees in these areas would require 20 years for re-establishment. In upland forest areas, many of which had been sprayed more than once, vegetative recovery was limited to the growth of bamboo and other undestroyed trees rather than to refoliation by the original prominent trees. Grasses invading and flourishing in these areas may inhibit the successful reestablishment of tree seedlings.

*See Chapter 5 for a discussion of genes and genetic engineering.

Large-scale damage to rubber trees in plantations also occurred in Vietnam. Accidental damage from defoliation to plants far removed from target areas, including agricultural crops, has been documented. In the highlands region a deliberate attempt at crop destruction (so-called resource denial) was made in order to eliminate food sources believed to be used by the enemy.

The entire herbicide defoliation program affected an estimated 2.2 million hectares in Vietnam. A 1974 report of the United States National Academy of Sciences concluded that herbicide use in South Vietnam caused serious and extensive damage. It may take a century for the hardwood tropical forests and coastal mangroves to fully recover.

What is the Effect of Defoliants on Animals? Complete spraying of mangrove vegetation with herbicides has a severe effect upon the animals living there. During their inspection Orians and Pfeiffer did not see a single species of endemic non–fish-eating birds, even though most areas throughout the tropics are rich in such bird species. Fish-eating birds, they found, suffered less from the defoliation, but even these were fewer than expected. The degree of herbicide toxicity in animals has not been completely determined. There have been occasional reports, however, of sick and dying birds and mammals in forests following defoliation. Furthermore, Dow Chemical Company, the manufacturer of two types of defoliant, says that no grazing of animals should be allowed on treated areas for up to two years after treatment, and that some broadleaf crops may show damage three years after application. In addition, as Orians and Pfeiffer remind us, "We must not forget that habitat destruction, which defoliation regularly accomplishes, is in most cases the equivalent of death for animals."

Agent blue, used mainly to kill rice crops, can also be directly fatal to man. A dose as small as 30 ml can produce acute poisoning resulting in headache, vomiting, diarrhea, dizziness, convulsions, general paralysis, and death. Agent orange is suspected of causing birth defects in humans. Large-scale spraying of these agents in Southeast Asia was ordered stopped in 1970.

In addition to defoliants, other activities of the Vietnam war had far-reaching ecological effects. Frequent air raids by bombers were estimated to have pockmarked the landscape with 2.6 million craters in 1968 alone. Altogether the bombing program left about 23 million craters averaging 7.5 meters in diameter and 12 meters deep, and destroyed 10 per cent of South Vietnam's rice lands. Such craters may form potential breeding grounds for mosquitoes, and they render agricultural areas useless (Fig. 6–5). Also, giant "Rome plows" were used to clear more than 200,000 hectares of vegetation. The plows

Figure 6-5 (Courtesy of Gordon Orians, University of Washington, Seattle.)

stripped the land at the rate of 400 hectares a day, destroyed wildlife habitats, and left the land open to erosion and leaching of mineral nutrients.

Ecocide, or ecological warfare, has included attempts at initiating large-scale forest fires. In Vietnam, the United States carried out at least three experiments aimed at causing major forest fires in order to remove forest canopy and concealment from the enemy. In one project, United States planes flew 225 sorties and dropped more than 250,000 gallons of herbicide over selected target areas. After defoliation, dead leaves were allowed to dry and then combustion was attempted by dropping incendiary bombs. Because of the prevailing humidity of the jungle forests, however, these experiments were not very successful.

Geophysical Warfare

Militarily applied technology has led to what has been termed *geophysical warfare*. Reports such as the Pentagon Papers describe operations in Vietnam in which cloud seeding by United States planes was used to produce rainfall in desired areas. Proponents of such types of warfare have considered the possibilities of steering of storms, manipulation of climate, and even inducing of earthquakes. Among the opponents to such tactics is United States Senator Claiborne Pell (Democrat, Rhode Island), who has proposed a draft treaty banning geophysical warfare, and many scientists have urged the formation of an international treaty to ban such warfare.

A major socioecological effect of the Vietnam war was the rapid rate of urbanization which occurred as people fled from the wartorn countryside to the city. Within one decade alone, the city of Saigon grew from a small town of 250,000 people to an overcrowded urban area of 3 million inhabitants. In short, the large-scale use of modern weapons in a small area has disastrous psychological as well as ecological effects.

In describing the ecological effects of these modern technological weapons in warfare Senator Gaylord Nelson (Democrat, Wisconsin) says, "A scorched earth policy has been a tactic of warfare throughout history, but never before has land been so massively altered and mutilated that vast areas can never be used again or even inhabited by man or animal."

ELECTRO-OPTICAL WARFARE

One new technology finding increasing use in weapons development is the laser. In the laser, high intensity light is projected in parallel waves at a single frequency so that the waves are in phase; that is, the trough and the peak of each reinforces the other to produce an extremely high energy beam of light (see Chap. 3). So-called *smart bombs* have been developed in which a laser beam is projected at a target by an attacking aircraft (not necessarily the plane that drops the bomb). The smart bomb dropped from the plane homes in on the reflected light from the laser beam. Proponents of such a weapon point to its increased accuracy, which reduces civilian casualties.

In 1972 the United States spent about $90 million on electro-optical warfare development. Laser weapon research is carried on in the United States in secrecy at the Kirtland airforce base near Albuquerque, New Mexico. There, one research objective aims at developing a laser beam which can be used to destroy enemy ICBMs. With such research the Buck Rogers ray gun may soon become a reality. Laser beams of extremely high energy can be projected over long distances and ignite blocks of wood a mile away. The laser gun has advantages—it is a straight beam traveling at 186,000 miles per second, and unlike conventional guns is unaffected by wind or gravity; thus a target five miles away, once in sight, would be essentially destroyed.

Robot aircraft, called RPVs (remotely piloted vehicles), represent another technology under development. In 1972, the United States Air Force spent about $100 million on the development and production of RPVs. The advantages of robot aircraft compared to manned vehicles are (1) the elimination of risk to human pilots; (2) their cheaper construction (because they do not require life support systems); and (3) the fact that they can maneuver with extremely sharp turns at g-forces that would cause unconsciousness in humans. One military spokesman has said, "It will be a great day when only machines make war and people make love."

THE COST OF MODERN WARFARE

In 35 centuries of recorded history mankind has averaged only one year out of 15 without war, and today there are more than 22 million men in the world under arms. Nearly everyone agrees that "war is hell," but there are some who believe it to be a necessary ingredient in the progress of civilization. British sociologist Stanislav Andreski believes that war has contributed in several ways to the advancement of civilization. In his view, war, with its accompanying necessity of social mobilization, has been a major factor in stimulating the processes of industrialization, the development of civil governments and nation states, and even culture and the arts. Many analysts believe that wars will continue at least for the foreseeable future. Indeed, war has become even more frequent and more devastating in recent times. Estimates indicate that since 1900 approximately 100 million men have died in war, whereas in the nineteenth century only about 3,845,000 were killed in war. Present nuclear weapons arsenals set for immediate retaliation pose a constant threat of accidental war.

George Rathjens examined the nature of the forces that foster an arms race. He found a sort of action-reaction phenomenon to be one of the major stimulants in the strategic arms race. As one side develops a technologically superior arms capability, the other side reacts with the same or even superior capability, which in turn causes another counter-reaction (Fig. 6–6). Rathjens suggested that a reduction in the uncertainty about the adversaries' intentions and capabilities is a necessary prerequisite to curtailing the arms race. Besides the gathering of intelligence, there are several other ways in which this reduction in uncertainty can be accomplished, including the unilateral disclosure of one's

Figure 6-6 By permission of John Hart and Field Enterprises Inc.

weapons capabilities, and disarmament negotiations. Many experts view the development of new technological weapons systems as merely a futile search for technical solutions to a political problem, namely, the problem of national security.

A tremendous burden of expense is placed on the people of the world to maintain modern weapons. In 1967, for example, world military expenditures totaled $182 billion — an average of $53 per year for every man, woman, and child on the earth. This amount of money is hard to comprehend, but it far exceeds the present cost of all United States lands, buildings, machinery, and cash. During the period 1964 to 1967, military spending increased more rapidly than the growth of the world population and more rapidly than the growth of the gross world product (Fig. 6–7). Thus, the burden of military spending is becoming greater, both overall and

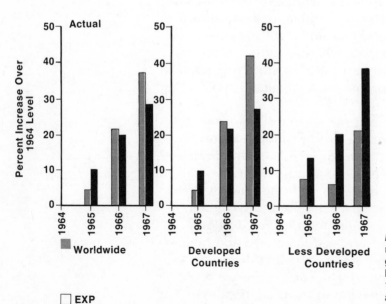

Figure **6–7** Four-year trend in military expenditures (EXP) and gross national product (GNP). The base year, with an index figure of 100, is 1964. (Adapted from Alexander, A.S.: The cost of world armaments. Sci. Am. *221*(4):21–27, 1969.)

per person. The world as a whole spends about 40 per cent more on military programs than on public education, and world expenditures on public health in 1966 were only about one-third as large as military outlays. Military spending imposes a particularly heavy burden on the underdeveloped countries, which can least afford such expenses. In the developed countries between 1964 and 1969, military spending rose twice as rapidly as gross national products, indicating the growing proportion of total national expenditure for weaponry, armed forces, and military engagements.

By applying modern technology to develop weapons man has purchased the threat of world destruction. As Robert McNamara, United States secretary of defense from 1961 to 1968 said:

> Technology has now circumscribed us all with a conceivable horizon of horror that could dwarf any catastrophe that has befallen man in his more than a million years on Earth.

Since the technological developments leading to the production of nuclear weapons, man has slowly moved toward limitations of their use. A partial nuclear test ban treaty was agreed to in 1963; a nuclear nonproliferation treaty went into effect in 1970; and the United States and Soviet Union since then have engaged in strategic arms limitations talks. Thus, we can hope that man will never face the horizon of horror that technology has given him the ability to produce. Konrad Lorenz writes in *On Aggression*:

> An unprejudiced observer from another planet, looking upon man as he is today, in his hand the atom bomb, the product of his intelligence, in his heart the aggressive drive from his anthropoid ancestors, which this same intelligence cannot control, would not prophecy long life for the species.

REFERENCES

Alexander, A. S.: The cost of world armaments. Sci. Am. *221*(4):21–27, 1969. (Survey shows a continuing world increase in per capita military expenditures.)

Case for war. Time, March 9, 1970, pp. 46–47. (Brief description of Stanislav Andreski's thesis: war was an important catalyst in the development of civilization.)

†Crichton, M.: The Andromeda Strain. New York, Dell Publishing Co., 1969 (Paperback). (Fictional account of the dangerous search for a new biological weapon.)

Gillette, R.: Smart bombs: air warfare undergoes a reluctant revolution. Science *176*:1108–1109, 1972.

Gilula, M. F., and D. N. Daniels: Violence and man's struggle to adapt. Science *164*:396–404, 1969. (Describes three theories of aggression and relates them to adaptation; suggests ways of diminishing violence.)

Goldblat, J.: Are tear gas and herbicides permitted weapons? Bulletin of the Atomic

†Fiction

Scientists 26(4):13–16, 1970. (Analysis of international agreements forbidding the use of CBW agents.)

Hirdman, S.: Weapons in the deep sea. Environment 13(3):28–42, 1971.

Johnson, R. N.: Aggression in Man and Animals. Philadelphia, W. B. Saunders Co., 1972 (paperback). (Probably the best single review of the subject, including theory and determinants of aggression and violence in society.)

*Jungk, R.: Brighter Than a Thousand Suns: a Personal History of the Atomic Scientists. London, Penguin Books, 1958 (paperback). (An exciting account of the early basic research and later application to the making of the bomb.)

Khan, H.: On Thermonuclear War. Princeton, New Jersey, Princeton University Press, 1961. (The nature, feasibility, and terrifying prospects.)

Langer, E.: Chemical and biological warfare (I): the research program. Science 155:174–179, 1967. (Description of active US program of 1967.)

Lorenz, K.: On Aggression. New York, Bantam Books, 1966 (paperback). (Classic work by an animal behaviorist including discussion of the question "What is aggression good for?")

Megargee, E. I., and J. E. Hokanson, eds.: The Dynamics of Aggression: Individual, Group, and International Analyses. New York, Harper & Row, 1970. (Collected reports by the original investigators.)

Meselson, M. S.: Chemical and biological weapons. Sci. Am. 222:(5):15–25, 1970. (Excellent review of the nature of CBW.)

Montgomery, B. L.: A History of Warfare. New York, World Publishing Co., 1968. (Illustrated account from ancient times to the nuclear age.)

Neilands, J. B.: Survey of chemical and related weapons of war. Naturwissenschaften 60:177–183, 1973. (Review of nature and effects of chemical agents of war.)

Neilands, J. B., et al.: Harvest of Death—Chemical Warfare in Vietnam and Cambodia. New York, Macmillan Publishing Co., Inc., 1972.

*Orians, G. H., and E. W. Pfeiffer: Ecological effects of the war in Vietnam. Science 168:544–554, 1970. (Firsthand description of effects of defoliation, bombing, and other military activity.)

Payne-Galwey, R.: The Crossbow. London, Holland Press, 1958. (Illustrated history of development of bows, crossbows, ballistas, and catapults.)

Rathjens, G. W.: The dynamics of the arms race. Sci. Am. 220(4):15–25, 1969. (An action-reaction phenomenon is a major stimulant in the strategic arms race.)

Storr, A.: Human Aggression. New York, Atheneum Publishers, 1968. (Is aggression an instinct?; and other topics.)

Thompkins, J. S.: The Weapons of World War III. New York, Doubleday and Co., 1966.

*U.N. Report No. E. 691, 24: Chemical and Bacteriological (biological) weapons and the effects of their possible use. New York, Ballantine Books, 1970 (paperback). (History of development, present state of technology, and effects of use of biological and chemical warfare agents.)

Westing, A. H.: Ecological effects of military defoliation on the forests of South Vietnam. Bioscience 21(17):893–898, 1971. (The extent of herbicide spraying and its ecological and economic consequences.)

Westing, A. H., and E. W. Pfeiffer: The cratering of Indochina. Sci. Am. 226(5):21–29, 1972. (Effects of saturation bombing and mass bulldozing on land.)

Wilkinson, F.: Edged Weapons. New York, Doubleday and Co., 1970. (Illustrated history of edged weapons, swords, crossbows, etc.)

Wilson, A.: The Bomb and the Computer. New York, Delacorte Press, 1968. (Wargaming from ancient Chinese mapboards to atomic computers.)

York, H. F.: Military technology and national security. Sci. Am. 221(2):17–29, 1969. (Antiballistic missile development as another example of searching for technological solutions to political problems.)

*Recommended further reading

PROBLEMS

1. Without aggression, would we lose our "spirit" and will to achievement?

2. Has aggression contributed to the advancement of civilization?

3. If we have learned aggression, do you believe we can learn to rid ourselves of aggression?

4. Is aggression in man an inherited instinct, a learned behavior, a result of frustration, or some combination of these? Support your arguments.

5. Select a series of random numbers from a random number table (a telephone book will substitute adequately for this problem). Let the first digit represent a television channel, second digit the day of the week, and the next three or four digits the hour of the day. (Some samples will not be practical hours and will have to be rejected.) Tune in your TV set at these times and channels and count the incidence of different types of violence for a period of one half-hour. After 10 samples or more, calculate the average number of incidences of violence per hour. Do you think this violence would have any effect on the behavior of viewers?

6. Do you think the scientists who developed the atomic bomb were morally responsible for its use?

7. Do you think there are instances when classified research (e.g., military weapons research) should be kept secret from the public? Why or why not?

8. It has been said by some that nonlethal "incapacitating" agents (e.g., mild germs or tear gas) should not be banned from warfare because they save lives and are more humane weapons. Do you agree or disagree?

9. Chapter 4 describes centers in the brain that control aggression by chemical injections. Do you believe that such control should be attempted? Why or why not?

7

ECOLOGY, POLLUTION, AND THE ENVIRONMENT

Freedom in a commons brings ruin to all.

Garrett Hardin

We are all are travelers on a spaceship. Our spaceship, called earth, is only the fifth largest planet revolving about a medium-bright star—the sun. It is the only planet known to be inhabited in this solar system. As space travelers, our survival depends upon the continuing function of the vital life support systems of the spaceship—the air, the water, and the soil. We are rather inexperienced space travelers, for we have not yet learned to properly care for and maintain our spaceship. We neglectfully dump poisons into our spaceship at a faster rate than our support systems can handle them. The danger now looms ahead that we may even bring these systems to the point of collapse. Our survival in the coming decades will depend, in part, on our ability to regulate and control the freedom of each individual to pollute our common spaceship earth.

WHAT IS AN ECOSYSTEM?

Recently the public has come to recognize the importance of ecology in understanding the environmental problems produced by technology; but few understand the basic principles

183

of this relatively new organized body of knowledge that now forms a familiar part of our vocabulary. Thus, before examining the effects of man's technological activities on the environment, let us first examine what we mean by the words *ecology* and *ecosystem.*

Ecology, derived from the Greek *oikos* meaning "house" or "place to live," designates that branch of science which studies the habitat or "house" of living organisms. It is often defined as the study of the relation of organisms or groups of organisms to their environment. Thus, it includes the study of pathways whereby energy flows between groups of organisms as well as the study of the cyclic flow of materials, chemicals, and nutrients between groups of organisms. It includes the study of populations, species, groups of individuals, and communities (all the populations occupying a given area). On a larger scale, ecology may include the study of entire communities and their relationship to the nonliving physical environment and how these function together as an ecological system or *ecosystem.* That portion of the earth which contains living organisms is usually designated as the *biosphere.* An ecosystem, then, is an area of nature that includes living organisms and nonliving substances, and includes energy interacting to produce an exchange of materials between these two groups. Man's attitudes toward the environment and growing population, coupled with technological advances, influences these ecological systems. In order to determine man's effect on natural ecosystems, ecologists first gather basic data on how these ecosystems function. Only by understanding how an ecosystem functions can we assess the impact of technology or pollution on that ecosystem.

A characteristic of any ecosystem is its *food web.* In order to understand how materials are channeled through an ecosystem, we must know which organisms are feeding on which other organisms. In general, we can define three basic feeding levels or *trophic levels:* (1) primary producers (plants), which convert carbon dioxide and light energy into plant material; (2) herbivores, which feed or graze on the plants; and (3) carnivores, which feed on the herbivorous animals. There are many variations on this simple scheme. In addition to the producers (green plants) and consumers (animals), ecosystems also usually include decomposers or reducers (bacteria and fungi). The biological activity of bacteria and fungi reduces dead plant and animal material to simpler chemical compounds, which enter the environment and are recycled and used again by the green plants (Fig. 7–1). Also, an ecosystem may contain many insectivores (organisms which feed on insects), secondary carnivores that feed on other carnivores, or omnivores, such as man, that feed on both plants and animals. Because an individual generally feeds on more than one type of plant or animal,

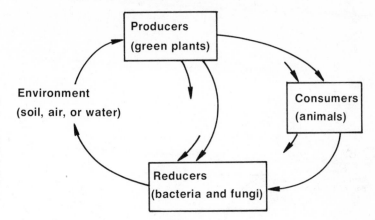

Figure 7–1 Circulation of materials between environment and organisms in an ecosystem,

In reality no ecosystem is isolated; each system would have exchange of materials with other ecosystems. (From Whittaker, R. H.: Communities and Ecosystems. New York, Macmillan Publishing Co., Inc., 1970.)

feeding relationships may form complex interconnections and result in a food web (Figure 7–2).

One important generalization that seems to be emerging from recent ecological study is that diversity of species in an ecosystem leads to stability of the ecosystem. For example, if instead of the complex interconnected relationships depicted in Figure 7–2, the entire ecosystem consisted only of crickets feeding on marsh plants and redwing blackbirds feeding on crickets, a small change in the environment could have serious effects. Eliminating redwing blackbirds, for example by hunting, would decrease the "grazing pressure" on crickets, which in turn would multiply and by their feeding reduce the level of marsh plants. Finally, without marsh plants the large cricket population would die down until it reached a level that could be sustained by the smaller available food source of marsh plants. Thus, when such a simple feeding system is disturbed, large and often undesirable fluctuations are likely to take place before some type of equilibrium is reestablished.

In a complex ecosystem, a change in the environment or the removal of one organism from the system may not introduce such large fluctuations in numbers of other organisms: a complex ecosystem is more stable over time than a simple ecosystem. Feeding relationships in most ecosystems are much more complex than the simple marsh plant–cricket–redwing blackbird interaction, and the effects of removing one organism or link in the chain may not be as predictable. Figure 7–1 shows that mergansers, green herons, and ospreys all feed to some extent on eels. Removing ospreys from this food web might cause an increase in the number of eels, thus supplying more food for herons and mergansers and possibly leading to an increase in their populations. Similarly, removing one species of tree from a tropical rain forest composed of hundreds of species might have little overall effect on the forest ecosystem, whereas removal of one species from a northern coniferous forest, where tree diversity is low, could have devastating effects.

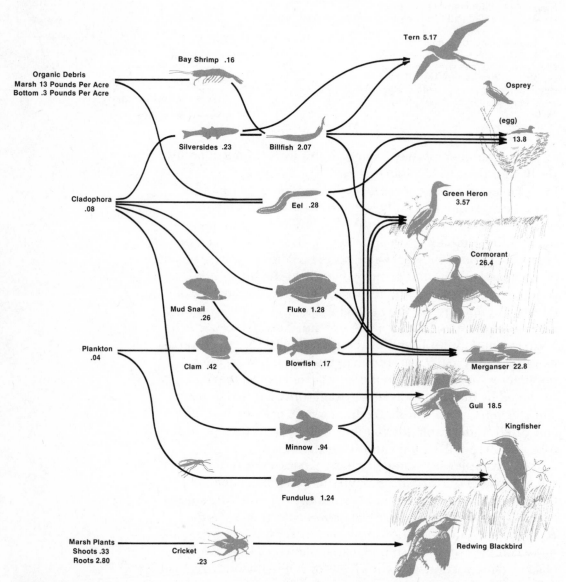

Figure 7–2 Portion of a food web in a Long Island estuary. Arrows indicate flow of energy. Numbers are the parts per million of DDT found in each kind of organism. (After Woodwell, "Toxic Substances in Ecological Cycles." Copyright © 1967 by Scientific American, Inc. All rights reserved.) (From Ehrlich, P. R., and A. H. Ehrlich: Population, Resources, Environment: Issues in Human Ecology. Second Edition. San Francisco, W. H. Freeman and Company, 1972. Copyright © 1972.)

In addition to the qualitative relationships of the food web (what is feeding on what), it is important to know the quantitative relationships between feeding levels. In examining these quantitative relationships, ecologists use measurements such as standing crop and productivity. *Standing crop* is generally defined as either the number of organisms or the biomass (weight of organisms) within a certain space. For example, we can count the number of rice plants per acre or determine the average number of kg of rice per square meter (biomass). By measuring the total number or weight of organisms per unit space for each feeding level we can construct community pyramids based on the standing crop at each trophic level — producer, herbivore, and carnivore. A more informative measure of community structure is *productivity* — the biomass produced per unit of space per unit of time. Rather than the number of rice plants per acre we would prefer to know the number of pounds of rice produced per acre per year; or in an aquatic ecosystem, the number of grams of fish produced per cubic meter of water per day.

The ultimate source of energy for all life on earth is sunlight (solar radiation). Plants, by trapping this energy through photosynthesis, form the basis of all food webs and are therefore called the *primary producers*. Only about 0.1 per cent of the solar energy that reaches a plant is fixed by photosynthesis. This solar energy is used by the plant to convert water and atmospheric carbon dioxide into chemical building blocks and eventually into carbohydrates (e.g., sugar or starch) and other plant storage and growth materials; during this process, oxygen and small quantities of water are released. On a worldwide scale about 150 to 200 billion tons of dried organic matter per year are produced by green plants. About ⅓ to ½ of this total is produced in the ocean. More than half of this photosynthetically fixed energy is used by plants for their own respiration. This energy used for maintenance and growth is not available for transfer to the next feeding level (herbivores). The energy that is transferred to the herbivores is also used, in part, for their own respiration. On the average, only 10 to 20 per cent of the energy in each trophic level is transferred to the next level. This loss of energy between trophic levels can be represented graphically as a community energy pyramid (Fig. 7–3*B*). Thus, quantitative measurements often indicate a reduction in numbers, biomass, or production between trophic levels (Figs. 7–3*A* and *B*).

Charting the flow of energy through an ecosystem from sunlight to plants to consumers to decomposers is a difficult task, but a few detailed quantitative studies have documented how energy flows through an ecosystem. For example, John Teal of the Woods Hole Oceanographic Institute has depicted

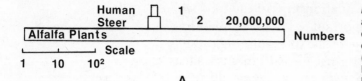

Human	1
Steer	2	20,000,000
Alfalfa Plants		**Numbers**

Scale

1   10   10²

A

Figure 7–3 *A,* Community pyramids for a shallow experimental pond of low nutrient content. Widths of steps for numbers of organisms are on a logarithmic scale. (From Whittaker, R. H.: Communities and Ecosystems. Macmillan Publishing Co., Inc., 1970.)

150 lbs. human

3,300 lbs. beef

27,000 lbs. alfalfa

B

B, The three types of ecological pyramid illustrated for a hypothetical alfalfa-calf-boy food chain computed on the basis of 10 acres and one year and plotted on a log scale. (From Turk, A., et al.: Environmental Science. Philadelphia, W. B. Saunders Co., 1974.)

(Figure 7–3 continued on the opposite page.)

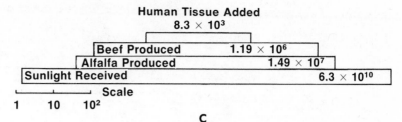

C, Food pyramid. The mass shown in each box is the amount required to produce the mass of tissue in the box above it. The areas in the boxes are proportional to the masses.

the food web and energy flow through the organisms of a Georgia salt marsh. The food web of this ecosystem (Fig. 7–4*A*) is composed of two types of photosynthetic plants, *Spartina* (a), a marsh plant, and algae (b) living in the water and mud bottom. The herbivores (c) consist of a group of insects that feed directly on *Spartina* plants and of a group of organisms (e), crabs, worms, snails, and so forth, that live at the level of the mud surface and feed on detritus, dead organic matter formed by bacterial decay of the plants. At the carnivore level, one group (d), including spiders, wrens, and nesting sparrows, feeds on the herbivorous insects, while another group (f), composed of mud crabs, raccoons, and snails, feeds on the detritus-feeding organisms of group (e).

This salt marsh ecosystem receives 600,000 kilocalories per square meter per year (kcal/m²/yr) of incoming solar energy. Most (93.9 per cent) of this energy is immediately lost from the ecosystem as back radiation and heat. Only 6.1 per cent of the solar energy is trapped by photosynthesis and used to reduce atmospheric carbon dioxide to organic plant tissue by *Spartina* (34,580 kcal) and algae (1800 kcal). Significant amounts of energy are again lost to the environment, this time through the respiration of *Spartina* (27,995 kcal) and algae (180 kcal). Of the remaining *net* photosynthetic production of *Spartina* (6585 kcal), 305 kcal are transferred to *Spartina*-feeding insects, and 3890 kcal are lost to the environment through the respiration of bacteria during decomposition of *Spartina*. Transfer of energy continues through the ecosystem to other members of the food web, and at each step some energy is lost to the environment through respiration. Examining the total energy budget of the ecosystem, 93.9 per cent of incoming solar energy never enters the food web; 6.1 per cent is converted to organic plant tissue, and 5.4 per cent of this is lost to the environment through the respiration of the plants, animals, and bacteria. Six-tenths of 1 per cent (6671 kcal) is left over as "net ecosystem production," which may be lost from the salt marsh

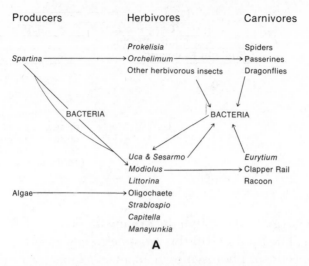

A

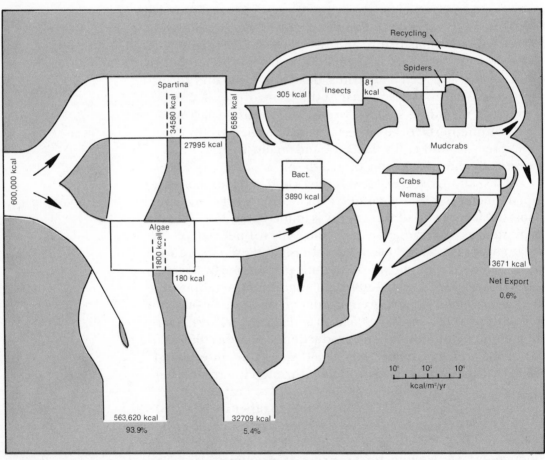

B

Figure 7–4 A, Food web of a Georgia salt marsh with groups listed in their approximate order of importance.
B, Energy-flow diagram for a Georgia salt marsh. (From Teal, J.: Energy flow in the salt marsh ecosystem of Georgia. Ecology *43*(4): 614–624, 1962.)

and find its way into adjacent or connected ecosystems of land
or water.

WHAT IS POLLUTION?

With the coming of the industrial revolution only 150 years
ago, man's capacity for modifying ecosystems took a quantum
leap. Man now diverts huge amounts of the energy flowing
through ecosystems toward his own use, and the by-products of
his industry are introduced into natural ecosystems, often with
disastrous results. Fossil fuels, once buried under the crust of
the earth, are extracted, refined, and made into synthetic (often
toxic) products that are released back into the biosphere. As industrial production continues to grow, the important question
to be decided within the next decade may be, as G. W. Woodwell has said, "How much of the energy that runs the biosphere
can be diverted to the support of a single species: man?"

Pollution, although difficult to define, may be considered
as any significant degradation or simplification of the food web
and energy exchange in a natural ecosystem or "the presence of
anything in undesirably large concentrations." This latter definition by ecologist Willis Hayes takes into account the fact that
some polluting agents are harmless or even beneficial in small
quantities. Pollutants are classified in two general categories: (1)
biodegradable organic wastes or nutrients; and (2) nondegradable harmful or toxic materials or energy. Both types produce
environmental stress and alter the energy flow available for
ecosystem production. An input of toxic materials which reduces the health of organisms will reduce ecosystem energy
production directly until organisms die or the entire ecosystem
becomes abiotic (without life). An input of usable energy, or
materials (such as biodegradable organic compounds), or nutrients (fertilizers) may stimulate ecosystem production at moderate levels (see *eutrophication,* p. 193), but higher inputs can
result in dangerous oscillations and finally lethal conditions
(Fig. 7–5). We shall consider some examples of these two general types of pollution.

BIODEGRADABLE WASTES OR NUTRIENTS

Sewage and Industrial and Agricultural Wastes

Throughout the world, especially in burgeoning urban
population centers, man must face the problem of disposing of
his own sewage waste while minimizing the harmful effects this
may have on the environment. Human or animal sewage and

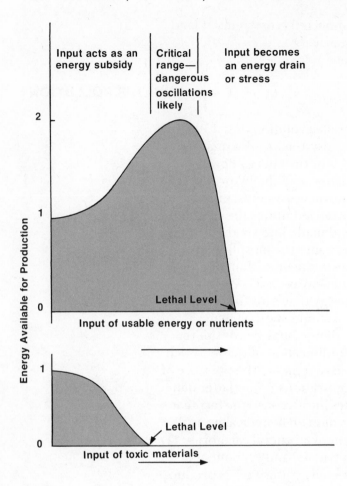

Figure 7-5 Schematic model of the effects of the two types of pollution—degradable organic (*upper graph*) and nondegradable toxic (*lower graph*). (From Odum, E. P.: Fundamentals of Ecology. Third edition. Philadelphia, W. B. Saunders Co., 1971.)

detergents are usually disposed of in some convenient body of water (river, lake, or ocean). This waste mixture contains high concentrations of organic matter and nitrogen, phosphorus, and other inorganic chemicals. Agriculture contributes large quantities of sewage material to aquatic ecosystems in the form of animal manure from domestic animal feed lots, poultry plants, livestock barns, food processing wastes, and fertilizers. Pulp, paper, and textile mills must dispose of large amounts of organic and inorganic waste.

If dumped directly into an aquatic environment, large amounts of organic matter may lead to a bloom of bacteria that derive their energy from decomposing organic material to simpler molecules. During the decomposition process bacteria utilize dissolved oxygen from the surrounding water.* Thus, a heavy bloom of bacteria can remove most of the oxygen from the water so that other oxygen-requiring organisms such as fish cannot survive. Consequently, in *developed* (industrial) coun-

---

*The amount of oxygen required to oxidize organic material to simple inorganic molecules is called the *biochemical oxygen demand* (BOD).

tries, sewage and other organic wastes often undergo *primary* and sometimes *secondary treatment* before being dumped into aquatic ecosystems. Primary treatment involves screening and sedimentation of solids. These are burned, buried, or rarely (but more desirably) made into compost for plant fertilizer. Secondary treatment involves aeration and a controlled decomposition of the organic "sludge" by microorganisms.

However, removal of organic matter alone from wastes does not eliminate its threat to the aquatic environment. After primary and secondary treatment, inorganic compounds such as nitrogen, phosphorus, and iron remain. When dumped into bodies of water, they can act as nutrients to promote (either singly or in combination) the growth of algae and other aquatic plants—a process called *eutrophication.*

Eutrophication simplifies an aquatic ecosystem to fewer species (Fig. 7–6). The resulting highly unsteady state may cause severe oscillations. In some bodies of water, after an initial algal bloom, nutrients become depleted, the algal population dies rapidly, and the dead algae provide organic material for a massive growth of bacteria, which in turn depletes oxygen from the water and leads to the death of oxygen-requiring animals.

Removal of inorganic as well as organic compounds from waste-water and elimination of the possibility of eutrophication is called *tertiary treatment.* Tertiary treatment can even produce recycled water for drinking, but it requires complex processes for the chemical removal of compounds, and may involve 3 to 20 times the cost of primary and secondary treatment. Hence, at present, tertiary treatment is used in few localities. In fact, on

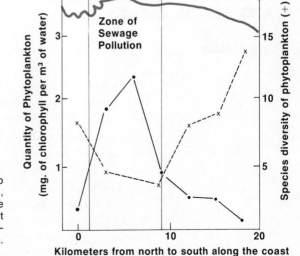

Figure 7–6 In a series of samples from north to south along the Mediterranean Coast of Lebanon, those within a zone receiving untreated sewage show a greater quantity of phytoplankton algae, but fewer species are present. • = quantity of phytoplankton; + = species diversity of phytoplankton. (From an unpublished study by the author.)

a worldwide scale most sewage waste-water now receives no treatment whatsoever.

Deforestation, Land Clearing, and Erosion

Man has undoubtedly had some influence on natural ecosystems for millions of years, but beginning with the agricultural revolution about 10,000 to 20,000 years ago, this influence has increased markedly. Man's activities in general have had the effect of simplifying ecosystems, thus making them in all probability more subject to disturbing fluctuations. Agricultural systems are one example of simple ecosystems produced by the activities of man. They require constant care and attention to keep them in a stable state. In a sense they may be thought of as putting all the eggs in one basket; thus, ten acres planted in one crop such as cotton may be totally decimated by the introduction of a pest such as the boll weevil. Although agriculture contributes greatly to our material well-being, it has had the effect of simplifying ecosystems on a large scale (see for example, Turk et al., 1972, Chap. 2). Deforestation, overgrazing, and faulty irrigation practices have changed formerly fertile areas into deserts. For instance, an area in western India that was a rich jungle 2000 years ago, is now known as the Thar Desert. Many areas within the historical "fertile crescent" of the Middle East have been ruined over the past 2000 years by poor agricultural practices and overgrazing. The famous ancient Cedars of Lebanon were cut from the mountain forests and used in the construction of Phoenician, Egyptian, and Persian ships as well as Solomon's and other temples. With the forests removed, valuable topsoil erodes into the sea.

NONDEGRADABLE, HARMFUL, OR TOXIC MATERIALS OR ENERGY

Toxic Metals

Toxic metal pollutants released by man into the ecosphere include mercury, cadmium, arsenic, beryllium, and lead. In general, they persist in the environment for long periods of time and enter ecosystems where they become concentrated through the food chain. As examples, we shall consider pollution of the environment by lead and cadmium.

Lead. Lead, a highly toxic substance, has become widespread in the environment in recent years. The average person in the United States now contains a level of lead in his bloodstream one-fourth of that considered diagnostic for lead poisoning. The toxicity of lead is well documented. Its symptoms include liver and kidney damage and deterioration of the central nervous system, including the brain, finally resulting in death. It acts by blocking the production of enzymes for hemoglobin synthesis in red blood cells. With less active hemoglobin, the blood is able to carry less oxygen, thus cells and organs which require large amounts of oxygen (e.g., nervous tissue) begin to deteriorate. There is no threshold below which lead is not toxic. Even the smallest detectable increase in lead in the blood above average background levels causes a decrease in enzyme activity for hemoglobin synthesis.

Lead-based ceramic glazes have been used in pottery manufacture for thousands of years. Some historians even attribute the fall of the Roman Empire to the fact that the Romans stored their wine in large jugs (amphorae) containing lead and sometimes "sweetened" the wine with lead acetate. The danger of this type of lead poisoning has been reduced in recent years by firing pottery at higher temperatures or, an even safer method, by the use of lead-free glazes.

Until recently lead was added as a conditioner to paint. House painters in the early part of the century were exposed to high levels of lead, and it is estimated that in 1918, 40 per cent of all painters showed evidence of lead poisoning. Lead poisoning from paint continues to be a problem, especially in urban areas of the United States. In old housing developments where leaded paint and window putty were used, infants and young children are exposed to peeling and chipping lead paint and putty which they inadvertently consume by chewing on woodwork or windowsills. This has resulted in a major health problem in some urban areas. Between 1954 and 1967, 2038 children were treated for lead poisoning in New York City; 128 (6.3 per cent) of these children died. Dr. Joseph Cimino, the medical director of the New York City Poisoning Control Center, believes that low-level chronic lead poisoning may be responsible for the scholastic underachievement of children in many older urban areas.

An even greater source of environmental lead contamination arises from the automobile. Tetraethyl lead has been added to gasoline since 1923 to increase its octane rating and to make it burn more smoothly. In 1966, for example, 190,000 tons of lead were released into the atmosphere from motor vehicles in the United States. In 1968, 300,000 tons of lead were added to gasoline. City dust may contain a lead concentration of about 1 per cent or nearly equal to that of many lead ores. Lead released into the atmosphere when leaded gas is burned can now be found throughout the biosphere — even in the Greenland ice sheet. Ice cores removed from the Greenland ice sheet indicate that lead concentrations beginning about

2000 years ago at less than 0.001 parts per million (ppm)* increased primarily in the last 50 years to 0.2 parts per million. Atmospheric lead may reach high concentrations in downtown urban areas, especially during times of heavy traffic flow. Lead concentrations are often greater in the air near highways and freeways and diminish with increasing distance from the highway. The average concentration of lead in the bloodstream of United States inhabitants is considerably higher than that of people in countries where little automobile traffic exists, and lead levels in the blood of urban residents are reportedly higher than those of suburban residents (Fig. 7–7). Laboratory rats exposed to the same lead concentrations as the average person in the United States (10 to 11 μg/m³ in six hours) developed brain and metabolic deterioration. Lead can now be found in the air, in people, in plants, in seawater, and in the soil. Craig and Berlin concluded, "Atmospheric lead at levels now common in urban areas is a health hazard."

Cadmium. Cadmium, another example of a toxic metal that poisons the environment, acts as an inhibitor of three enzymes important for metabolism and respiration. Although its close chemical relative, mercury, has received considerable attention as an environmental pollutant, more than twice as much cadmium as mercury is used in industrial processes each year in the United States. There is increasing evidence that cad-

---

*One part per million refers to 1 volume of a gas or 1 weight of a particulate pollutant per million volumes of air or water.

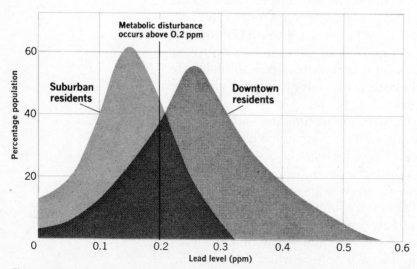

Figure 7–7 Lead levels differ in suburban and downtown inhabitants. (Reprinted with permission from Hall, S.: Pollution and poisoning. Environ. Sci. Tech. 6(1):31–34, 1972. Copyright by the American Chemical Society.)

mium pollution may pose serious health problems. A dramatic case of cadmium poisoning occurred in northern Japan in 1962. Two hundred twenty-three people living in an area exposed to high concentrations of mining waste became affected with a strange degenerative bone disease later identified as cadmium poisoning. The disease resulted in decalcification of the bones, which literally disintegrated, and the patients' bodies appeared transparent in x-ray photographs. Statistical surveys in the United States also indicate a *possible* correlation between airborne cadmium concentrations and deaths caused by high blood pressure and arteriosclerosis.

Where does cadmium come from? Cadmium is used in many industrial processes and forms a component of thousands of products. For example, rubber automobile tires contain 20 to 90 ppm of cadmium, and estimates indicate that during 1968 approximately 11,400 pounds of cadmium were released into the atmosphere by tire wear alone. About 90 per cent of the cadmium used in 1968 went into electroplating, paint pigments, plastics, metal alloys, and batteries. Significant amounts were also used as stabilizers in polyvinylchloride plastics. About 45 per cent of the total atmospheric emissions of cadmium in 1968 came from operations in which cadmium-bearing ores were processed, and about 52 per cent came from operations in the disposal or recovery melting of scrap metal containing cadmium.

How widespread is cadmium? Cadmium from such industrial processes releases about 4.6 million pounds into the atmosphere each year in the United States. In a 1970 survey, 42 per cent of 720 water samples tested from rivers and reservoirs in the United States contained significant amounts of cadmium. One study even found that some tap water had cadmium concentrations greater than the United States Public Health Service permissible levels. Cadmium-containing fertilizers and pesticides are another major source. Americans take in cadmium in their air, water, food, beverages, and cigarettes.

In some cases, the concentration in body tissues of Americans is greater than those which cause biochemical and liver changes, high mortality, and shortened life spans in experimental animals. Doctors at the Dartmouth Medical School have analyzed the concentration of cadmium in foods and found high concentrations in certain seafoods and in beef kidney. Studies on human populations in different countries have shown that cadmium concentrates particularly in the kidney; and cadmium levels in humans are highest in Japan, at an intermediate level in the United States, and lowest in Africa. As more studies are conducted cadmium may prove to be the most important of the toxic metal environmental pollutants.

Persistent Organic Chemicals

Many synthetic organic compounds represent significant environmental pollutants. These include plastic products, oil, pesticides, herbicides, some components of photochemical smog, and many other organic chemicals probably not yet identified as major pollutants. We will discuss briefly only one group of the lesser known organic pollutants, the polychlorinated biphenyls (PCBs).

PCBs are widely used industrial organic chemicals which can now be found in rainwater, in human tissue, and in many species of birds and fish. They are stable compounds, which tend to accumulate in food chains and are stored (like the pesticide DDT) in fatty tissue. How they became so widely distributed or what biological effects they may have is not completely understood. They are used industrially in dielectric fluids, in capacitors and transformers, and in hydraulic fluids, heat transfer fluids, plasticizers, solvents, adhesives, sealants, paints, and printing inks. Their use has grown steadily since 1930, and United States sales in 1970 reached 34,000 tons. The estimated cumulative production over the years has been about 400,000 tons. PCBs may enter aquatic environments through industrial disposal waterways and sewage outfalls. It is also estimated that in recent years about 1000 to 2000 tons of PCB have escaped into the atmosphere from the natural degradation of disposed plastic materials containing PCBs. Controversy now exists as to whether or not some PCBs are actually degradation products of DDT.

The lifetime of these chemicals is not definitely known but (like DDT) they are quite resistant to biological or physical decomposition. They are found in highest concentrations in polluted coastal waters, but even fish from "clean" arctic lakes may contain 0.01 to 1.0 ppm of PCBs. They appear to be concentrated in fish-eating birds.

Like DDT, PCBs concentrate in fatty body tissue and can now be found in human beings. In 600 samples of human fatty tissue, the United States Environmental Protection Agency (EPA) found 33 per cent to contain at least 1 ppm of PCB. One older man who had been continuously exposed to industrial processes using PCBs had more than 100 ppm in his fatty tissues. PCBs reach human beings through the food chain and are particularly concentrated in some fish, but the sources of PCBs are often difficult to identify. For example, in a child who was found to contain high concentrations of PCBs, investigators found that the PCBs originated from breakfast cereal contained in a paper box made of recycled paper. The paper used for recycling was originally printed with PCB-containing ink.

The biological action of PCB is not clearly understood, but

it is known to alter liver tissue and affect a variety of enzymes. In Japan 1000 people ate rice oil contaminated with PCBs leaked from the heat exchanger in a rice processing plant. These people developed clinical symptoms including darkened skin, eye discharge, and severe acne.

As one example of an organic chemical pollutant, PCBs demonstrate the fact that other organic chemicals should be examined as possible environmental pollutants. As A. L. Hammond has said, "Their presence and persistence (in the environment) reemphasize the likelihood that any widely used industrial chemical may become an environmental pollutant, and increase the responsibility for public disclosure of production quantities and use patterns . . ."

Heat

Temperature is one of the most important controlling factors governing the distribution, growth, and reproduction of organisms on the earth. Waste heat, a by-product of almost all industrial processes, constitutes a major environmental pollutant. Thermal pollution can be defined as any unwanted heat energy accumulating in any phase of the environment, or any temperature change which adversely affects water quality and the beneficial uses of water. The recognition of thermal pollution as an environmental problem occurred only recently, and studies on the subject before 1968 are conspicuously absent.

Large quantities of fresh or marine water are used in industrial cooling processes for temperature exchange and then released back into streams, lakes, or oceans, raising the temperature of the surrounding water. Such use of river, lake and estuarine waters for industrial cooling poses a threat to aquatic life in some areas. According to a 1966 report by the Northeast Illinois Planning Commission, in an industrial area like Chicago, about 15 times as much water is used for industrial cooling as for domestic purposes. Excluding irrigation, about 90 per cent of all water is used for industrial cooling or absorption of heat. Steam electric power plants, chemical industries, ironworks, and air conditioning units are at present the major consumers of water for cooling purposes. In 1968, during a United States Senate subcommittee hearing on air and water pollution, Senator Edmund Muskie discussed the nature of the thermal pollution problem. The committee estimated that by 1980, in some areas of the country during periods of low water flow, all available water from streams will have to be passed through power generating stations for cooling usage.

The principal contributor to thermal pollution is the electric power industry. Assuming about 1.26 million calories of heat rejected for each kilowatt hour of electricity generated, and a national United States production of 728 billion billion kilowatt hours, the total heat rejection to rivers has been estimated at 3.02×10^{17} calories. Demands for electrical energy are increasing rapidly, and the Federal Power Commission estimates that by 1980 at least three times the present quantity of electricity will be required. To meet this skyrocketing demand for energy many electrical generating plants are being planned or are currently under construction.

As atomic research progresses, large nuclear generated electric power plants are being designed. Nuclear power plants use heat from uranium fuel to produce steam, which drives turbine generators. This steam must be cooled and condensed before being returned to the reactor for another cycle. Water is drawn in from the environment to provide this cooling and then returned to its source carrying with it large quantities of heat (Figs. 7-8 and 7-9). Nuclear power plants are less efficient than conventional fossil fuel plants and produce 50 per cent more waste heat per unit of electricity generated. Thus, temperatures near the outfall at nuclear plants may be raised 10 to 20°C above their normal level, with detrimental effects on biological communities. The 16 nuclear power plants either currently operating or ordered for location on the shores of the Great Lakes would have met nearly all the United States demand for electricity in 1940. Their operation will release an amount of heated water equal to the total flow of the Mississippi River.

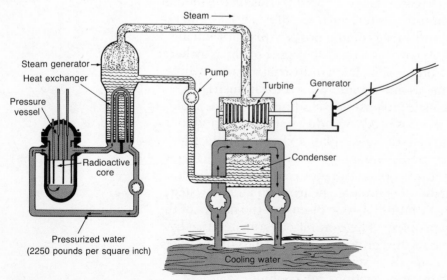

Figure 7-8 Schematic illustration of a nuclear power plant. Water is taken in from a stream, lake, or the ocean, used to absorb heat, and returned at a higher temperature to the original body of water. (From Turk, A., et al.: Ecology, Pollution, Environment. Philadelphia, W. B. Saunders Co., 1972.)

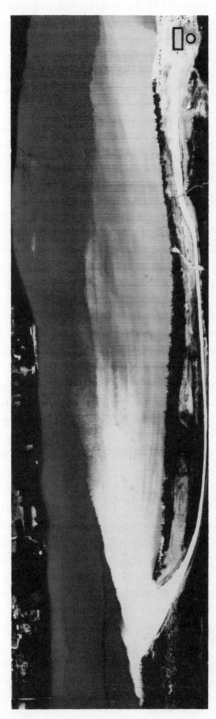

Figure 7-9 Nuclear power plant at Haddam on the Connecticut River empties up to 370,000 gallons of coolant water a minute through a discharge canal (*bottom*) into the river. In this aerial thermogram, made by HRB-Singer, Inc., for the U.S. Geological Survey's Water Resources Division, temperature is represented by shade of gray. The hot effluent (*white*) is at about 93 degrees F.; ambient river temperature (*dark gray*) is 77 degrees. The line across the thermogram is a time marker for a series of absolute measurements. (Courtesy of US Geological Survey.)

The dispersal of heat from water to the atmosphere depends upon the speed and turbulence of receiving water currents, the temperature difference between the water and the air, and the speed and direction of the wind. An increase in water temperature can result in certain physical and chemical changes in the water itself. It may affect the solution rate and solubility of minerals, precipitation of compounds, pH (acidity or alkalinity), and the solution of atmospheric gases, the most significant of which is oxygen. Dissolved oxygen in the water, which is important for fish and other aquatic life, decreases with a rise in temperature (Table 7–1). Also, if organic pollution is present, the increased biochemical oxygen demand (BOD) can further lower the oxygen concentration.

Discharged waste heat has significant biological effects on aquatic life. Studies on fish show that a rise in water temperature increases their oxygen consumption and swimming activity while decreasing their efficiency at converting food intake into body weight. Waters above 33° C are considered uninhabitable for most fishes, and still lower temperatures may be necessary for fish reproduction and normal egg development. In addition, increased temperature may shorten the life span of animals. Temperature conditions must remain favorable for all components of the environment including algae and other aquatic plants, small crustacean animals, and bait fishes, all of which constitute part of the food web. For example, raising estuarine water temperatures to 20° C may not immediately harm fish populations but will inhibit the reproduction of eel grass plants and ultimately will reduce the productivity of the estuary.

In the summer of 1968, during the testing of a new power plant on Cape Cod Canal, a large number of menhaden were trapped in the 34 to 35° C effluent and were killed. Fish kills resulting from thermal pollution are rare, but more subtle effects on the community structure such as reduction of diversity and perhaps of stability of the ecosystem are becoming apparent as more studies are completed. As with other forms of pollution, a synergistic effect may occur in which, for example, a combination of sewage pollution and thermal pollution may

TABLE 7–1 Solubility of Oxygen in Water at Different Temperatures

Temperature (°C)	Solubility (parts per million)
25	8.4
15	10.2
5	12.5

have a greater detrimental effect on a biological community than either form by itself.

J. R. Sylvester of the University of Washington College of Fisheries reviewed existing studies of the effects of thermal effluents on fish. He concluded that slight increases in temperature can be beneficial in stimulating feeding, growth, and overall general activity of fish, but when combined synergistically with other types of pollution, even minor temperature increases can be lethal. These "lethal limits" may be avoided by fish if they can migrate away from such an area. However, temperature increases of 10° C or more, while not fatal, can cause stress in some fish so that their activity is significantly affected; they will not necessarily reproduce, feed, or grow at such sublethal increased temperatures. Many desirable game fish require a water temperature below 10° C for reproduction.

F. J. Trembley studied the effects of thermal discharge from Delaware River power plants on populations of algae at, above, and below, the site of heated discharge. He found greater numbers of algae, especially blue-green algae, but fewer species at the discharge and as much as 650 feet downstream. Ruth Patrick found similar results by recording the growth of various types of algae in the Sabine River where, as the temperature was raised from 20 to 30° C to greater than 35° C, a change in the type of algal species growing on microscope slides occurred.

Heated water is not the only problem for aquatic life resulting from power generating plants. Nuclear plants may release certain radioactive isotopes into the water (discussed later in this chapter). Communities of organisms may also be affected by metals dissolved in effluent water, and by detergents, acids, and chlorine used to keep power plant condenser tubes free of fouling by biological growths and deposits. A study by Allen Brook and A. L. Baker of the University of Minnesota indicates that acid chlorination of river cooling water passing through a power plant depressed the rate of photosynthesis and respiration of the river phytoplankton to a much greater extent than heating (Fig. 7–10).

Two principal methods are used to reduce the temperature of cooling water before returning it to streams, lakes, and estuaries. In the first method of control, cooling water is distributed into large outdoor ponds where evaporative cooling releases the heat into the atmosphere. Unfortunately, the large land area required for cooling makes this method of control rather expensive. It is estimated that for a large power plant planned for the future a cooling lake of 1000 to 2000 acres (1 mile by 3 miles) would be required. In the second method of control, the heated water is pumped into the lower part of a 300 to 400 foot high "cooling tower" over baffles, and as the

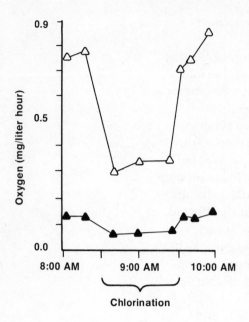

Figure 7-10 Sequential rates of phytoplankton photosynthesis (PS, open symbols) and respiration (R, closed symbols) of samples collected from power plant discharge before, during, and after chlorination. (From Brook, A. J., and A. L. Baker: Chlorination at power plants: impact on phytoplankton productivity. Science *176*:1414–1415, 1972. Copyright 1972 by the American Association for the Advancement of Science.)

warm air rises in the tower, the water cools by evaporation. However, cooling towers release large amounts of water into the atmosphere and may even affect weather by producing fogs or freezing mists during the winter in temperate climates. They are not effective for use with salt water because of the concentration of salt from the evaporated water. A third method, known as dry tower closed radiators, are also being used, but they are a very expensive means of water cooling.

Debate continues on the overall significance of thermal pollution. Some scientists have even suggested that heat energy derived from power generation could be usefully employed for man's benefit. For example, the waste heat could be used to keep navigable waters ice-free during the winter. Another proposed scheme involves pumping heated water into the depths of the ocean, where due to its lighter density, it would rise, bring nutrient-rich water to the surface, increase productivity, and result in greater production of fish protein for human consumption. A more practical use might be in providing warm water for controlled mariculture (see Chap. 9).

Some scientists believe that increasing thermal pollution poses a threat to all life on earth. Lamont Cole has discussed the long-term effects of thermal pollution on the earth as a whole. The large-scale use of fossil and nuclear fuels rapidly releases into the atmosphere energy that has been stored for millions of years. Making certain assumptions, Cole estimated the amount of increase in the earth's heat budget (incoming solar radiation minus outgoing back radiation) necessary to produce a given mean earth temperature. For example, a 4.2 per cent *increase* in the earth's heat budget would raise the average temperature of the earth by 3° C—enough to melt the polar ice caps. A 4.1 per cent *decrease* would bring on a new ice age. The quantity of energy now being produced by man is only 0.025 of 1 per cent of the total heat radiated by the earth. This has an insignificant direct effect on the average earth temperature; but what about the future? A 7 per cent per year increase in energy output (a current value), over the next 91 years would result in a warming of 1° C in the average earth temperature—enough to influence boundaries between plant communities. Following the present trend, it would take 780 years to melt the polar ice caps (flooding many large cities) and 980 years to raise the average temperature to 30° C, which would make the planet uninhabitable. Of course, mankind must solve the general problems of growth within a much shorter time span (perhaps a generation) if he is to survive as a species (Chap. 10).

The side effect of fuel combustion is the release of smoke and particulate matter, which absorb incoming radiation and have a cooling effect upon the earth, counteracting to a certain extent the release of heat. On the other hand, water vapor and carbon dioxide form an atmospheric layer transparent to sunlight but opaque to longer wavelength back radiation. This results in a so-called *greenhouse effect* and tends to raise the average surface temperature of the earth.

Data on past, present, and future trends in particulate matter production, CO_2 concentration, and heat load are not sufficient at present to allow us to predict accurately the ultimate consequences of these interactions on the global climate, and therefore we should not place great confidence in the exactness of Cole's predictions. However, it is precisely in this lack of confidence that the danger lies; Cole's figures may not be correct, but the general trend indicated by them cannot be ignored. Thermal pollution is but one example of mankind's global experiments in vast environmental modification—the results of such experiments (including proposed regional climate-modifying schemes) are difficult to predict, but they could be catastrophic.

Noise

One increasingly disturbing factor, especially in urban environments, is noise (unwanted sound). Noise resulting from technological development and crowding commonly disturbs our sleep, makes conversation difficult, may cause hypertension, and in some cases results in permanent hearing damage. One does not have to look far for evidence of noise pollution today.

The intensity of sound is measured in a scale of decibels

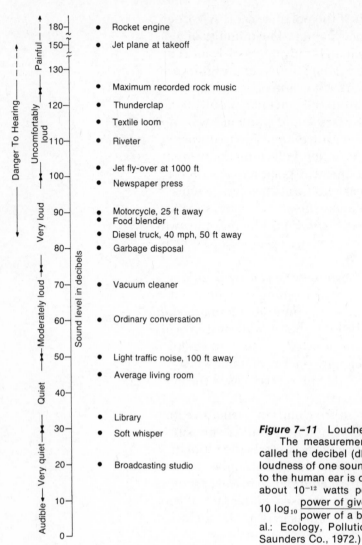

Figure 7–11 Loudness and the decibel scale.
The measurement of loudness commonly uses a unit called the decibel (dB), defined in terms of the ratio of the loudness of one sound to another. The softest sound audible to the human ear is defined as zero decibels and is equal to about 10^{-12} watts per m^2. Thus, loudness in decibels $=$ $10 \log_{10} \dfrac{\text{power of given sound}}{\text{power of a barely audible sound}}$. (From Turk, A., et al.: Ecology, Pollution, Environment. Philadelpha, W. B. Saunders Co., 1972.)

(dB) (Fig. 7–11). Hearing ability is measured in terms of the threshold decibel level at which a person can detect sounds at different frequencies.

Hearing often degenerates somewhat with age, particularly for the higher frequencies. Men in their 60s have, on the average, a hearing loss of about 30 dB at the higher frequencies; but repeated and continuous exposure to high noise levels can result in permanent hearing loss at earlier ages (Fig. 7–12). One study found that 400 men working in a noisy factory environment (where they were subjected to an average of 90 dB, between 100 and 6000 cycles per second for a period of 40 years) suffered substantial hearing loss, particularly at the higher frequencies. Even young men of 30 had so much impairment that they found it difficult to understand normal speech. Studies by the American Standards Association indicate that, during an eight-hour working day, people can tolerate about 85 dB of noise without substantial damage to hearing.

The basic hearing mechanism of the ear is the cochlea, a snail-shaped organ containing more than 20,000 small hair cells that transfer sound to the brain. Loud noises damage these cells. During quiet periods many of them can recover, but with long and protracted noise many are damaged beyond repair, and deafness can result. There is some evidence that noise can result in physiological stress; and sudden noise can temporarily raise blood pressure, quicken pulse rate, reduce spontaneous movement of the gastrointestinal tract, tense muscles, and constrict tiny blood vessels of the skin.

Reports indicate that between 6 and 16 million people in the United States alone are going deaf from job-related noise, and that $4 billion is lost by industry each year due to accidents, absenteeism, inefficiency, and compensation claims — all because of noise. In contrast to the large number of people with noise-related hearing loss in the United States, studies of remote primitive tribes in Africa (who live in very quiet environments) show that hearing remains excellent even in the elderly.

Where Does Noise Come From?   Noise pollution is nothing new, although its extent and magnitude are certainly increasing. In ancient Rome Julius Caesar was so provoked by the noise of chariots in the streets that he almost had them totally banned. In many areas environmental noise has doubled in the past ten years. People in urban environments are constantly bombarded with noise from construction projects, air and surface transportation, and industries and in many vacation retreats by minibikes, snowmobiles, and radios. Probably the largest, most widespread, and most annoying source of noise in the urban environment is motor vehicles. Trucks, buses, motorcycles, and sports cars are among the noisiest of these. It is estimated that motor vehicles account for an average of 75 per cent of the urban decibel count.

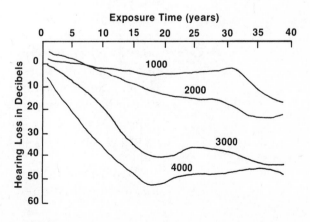

Figure 7–12 Cumulative deafness, due to years of exposure to a noise level averaging 90 decibels in each frequency band, shows two peaks. The maximum hearing loss at frequencies of 3000 cycles per second and above is reached in about 15 years. Loss of hearing at frequencies of 2000 cycles and below, however, does not reach its maximum until after 30 years or more. (Adapted from Beranek, L.: Noise. Sci. Am. *215*(6):66–76, 1966.)

Airplanes are another source of noise pollution, and people living near airports may be subjected to repeated high levels of noise during takeoffs and landings. The high noise area around New York's Kennedy Airport is 23 square miles — an area containing 35,000 dwellings, 108,000 residents, 22 public schools, and several dozen churches. The recent development of supersonic transport (SST) planes raises the possibility of sonic booms occurring over populated areas. Such airplanes, traveling faster than the speed of sound, produce shock waves that trail behind the plane and generate a cone-shaped pressure wave that constitutes the sonic boom. The path of the boom is about 50 miles wide when the plane is flying at an altitude of nine miles. Although the intensity of a sonic boom at ground level decreases as the plane flies higher, tests have shown that people within eight miles of the center line of a boom path are disturbed significantly by the booms.

The modern household kitchen can be a serious source of high noise levels. The combination of fan, dishwasher and garbage disposal can produce a total sound level of over 100 dB — a level of intensity which, if continuous, could cause permanent hearing damage from a full day's exposure.

Millions of industrial workers are employed at occupations in which noise levels caused by machinery are dangerously high and could result in definite hearing loss. For example, rock drillers working in a quarry may experience noise levels of 100 dB all day even though noise at this level is safe for only five minutes per day.

What Can Be Done to Control Noise Pollution?   Unlike some other environmental pollutants, the knowledge and technology to control noise pollution already exist. Engines can be more efficiently muffled, machines can be made to run more quietly, and buildings can be insulated and soundproofed against noise. Although efforts toward reducing noise pollution may be expensive, they are necessary. Methods for controlling and reducing noise include instituting and enforcing noise codes, improving the design and operation of quieter vehicles, the creation of buffer zones to separate residential areas from airports and superhighways, and better design of noise-insulated buildings. Some industries are attempting to protect their workers in noisy locations by requiring them to wear earplugs, earmuffs, or both, and attempts are being made to reduce noise from machinery. Several states have passed legislation outlining permissible and nonpermissible noise levels under certain circumstances. Much of the noise within buildings could be controlled by use of more solid and noise-insulating construction techniques.

Under the 1972 Noise Control Act, the EPA is responsible for the coordination of federal programs for noise research and control, for development of criteria to protect public health and welfare, and for setting standards for such processes as construction and transportation equipment, product labeling, research on noise effects, development of measurement and control techniques on noise, and technical assistance to the states. The Noise Control Act sets decibel standards on machinery and products known to be potential or actual noise hazards. Fines of up to $25,000 can be imposed on manufacturers who do not comply, and citizens may sue if the government does not act within 60 days of the suspected violation. The bill also provides for the use of government purchasing to pay up to 25 per cent over regular prices for low noise level equipment and machinery and offers technical assistance to set up local noise control programs.

Radioactivity

Discounting atmospheric nuclear bombing or full-scale nuclear war, the largest potential source of radioactive pollution is the growing number of nuclear power electrical generating plants. About 25 nuclear power plants have received operating licenses in the United States. Another 117 are planned or under construction. Commercial nuclear power generating plants use uranium pellets as fuel. These pellets are enclosed within vertical thin-walled tubes or rods of corrosion-resistant metal such as stainless steel or an alloy of zirconium. Nuclear fission from these uranium rods produces heat, which is used to heat water circulating through a boiler to drive steam turbines and generate electricity (see box insert).

HOW NUCLEAR POWERED ELECTRICAL GENERATING PLANTS WORK AND THE EFFECTS OF RADIATION DOSAGES ON ORGANISMS

Currently operating nuclear power plants depend upon fission. In a typical reactor, the uranium-235 fuel is bombarded with slow-speed neutrons. The U-235 nucleus splits into two smaller nuclei, releasing two or three neutrons, which fly off and split additional uranium atoms in a chain reaction producing large amounts of heat. The energy results from a conversion of mass into energy $(E=mc^2)$, because the total mass of the products of such reactions is found to be less than the mass of the reactants (see also box insert, Chap. 6).

Uranium pellets about the size of a cigarette filter are stacked into sealed tubes about 3 meters long. Many of these tubes are arranged within the walls of a large reactor. Among the fuel rods are dispersed graphite or boron "control rods" or "poison rods," which can absorb neutrons. When all control rods are inserted among the fuel rods, no reaction takes place.

The degree of insertion of the control rods determines how intense the reaction becomes and how much heat is generated.

Generally, the core of the reactor is surrounded by water, which acts as a moderator to slow down the escape of stray neutrons and to cool the reactor by absorbing excess heat. The circulating water jacket also serves to carry away heat from the reactor through a system of piping, where it is used to produce steam, which drives turbines to generate electricity. Large heat build-ups, if not removed, can result in a dangerous "meltdown" of fuel rods. The reactor is surrounded by several layers of concrete or steel to prevent the escape of the highly radioactive fission products.

Radioactive atoms decay as a result of the spontaneous emission from the nucleus of alpha, beta, or gamma rays. Alpha particles contain two protons and two neutrons; beta particles are electrons; and gamma rays are forms of high energy electromagnetic radiation. X-rays are electromagnetic waves like ordinary light, but of shorter wavelength and higher energy. When charged particles such as electrons strike heavy atoms of a target and cause the inner orbital electrons of those atoms to escape, the *return* of electrons to fill the vacancies in the inner energy levels produces x-rays. In addition to the radioactive products of man-induced nuclear reactions such as fission, many elements, especially the heavy ones, have naturally occurring radioactive isotopes which contribute, along with cosmic rays, to the natural background radiation on the earth's surface. Isotopes are defined as atoms of the same element (hence having the same number of protons), which differ in mass due to a different number of neutrons. Whenever an atom undergoes either alpha or beta decay, it changes its elemental identity. For example, when the radioactive carbon isotope carbon-14 undergoes beta decay it becomes the stable nitrogen isotope nitrogen-14. Decay of many of the heavier elements, such as the isotopes of uranium (which are all radioactive), results in the production of isotopes of other elements which are also radioactive, which finally decay through a number of radioactive intermediates to become a stable isotope of lead. Each radioactive isotope demonstrates a characteristic *half-life,* which is the time it takes for one-half of the atoms to decay. For example, strontium-90, with a half-life of 25 years will, after 25 years, emit only one-half its original intensity of radiation, by 50 years one-fourth, by 75 years one-eighth, and so forth.

Radiation dosage of organisms is measured in terms of units called *rads.* One rad is equal to the absorption of 100 ergs of energy per gram of tissue. Radiation damage depends upon the *dose rate* or number of rads received per unit time. Large single doses delivered over a short time period (minutes or hours) are known as *acute doses.* In contrast, *chronic doses* of sublethal radiation may be experienced over generations. Sensitivity to radiation differs both between groups of organisms and between the life stages of organisms. For example, an acute dose of 2000 rads that would kill mature mammals (such as man), in insects might affect only the more sensitive reproductive stages, and in microorganisms might have little or no effect. In humans, acute exposure to a dose of 400 rads results in 50 per cent population mortality (LD_{50}) and 650 rads results in nearly 100 per cent death in human populations. Much lower level sublethal doses can result in delayed symptoms including cancer, cataracts, increased genetic mutation, decreased fertility, and shortening of life span.

The fission products from uranium include such radioactive waste products as strontium-90, cesium-137, radioactive iron isotopes, and gaseous isotopes of iodine, krypton, and xenon. The tubes containing the uranium frequently develop tiny pinhole leaks or fractures, allowing the radioactive fission products to escape and contaminate the cooling water. These radioactive wastes, released in the liquid effluents, may be transmitted directly to humans through treated or untreated water supplies, or indirectly through crop irrigation, stock watering, recreational acitivities such as swimming, hunting, fishing, and through eating contaminated waterfowl or fish; or the suspended radioactivity may settle and accumulate in stream beds or be concentrated by aquatic organisms and subsequently transmitted to humans through consumption of fish or shellfish. Some gaseous isotopes may be released through the power plants' stack effluent, directly exposing persons in the vicinity to fallout and possible inhalation of radioactivity, or they may contaminate surrounding cisterns, water supplies, topsoil, or vegetation through the settling of radioactive particles.

Radioisotopes differ in their active half-lives and in their biological activity or tendency to concentrate in living organisms. Cesium-137 and iodine-131 tend to accumulate and become highly concentrated in certain plants and animals. The radioactive hydrogen isotope, tritium, occurs in reactor wastes as heavy water, and while not a highly hazardous material compared to other radioisotopes, it has a long half-life (12.5 years) and is produced in relatively large quantities in pressurized water reactors. No practical method for tritium containment exists, and some scientists have become concerned about an eventual tritium build-up and contamination of the world hydrosphere, particularly if fusion reactors become commonplace.

Fuel elements of the reactor must be *reprocessed* about every 6 months. During this reprocessing, uranium and plutonium are separated and returned to the reactor, and gaseous krypton-85 (half-life 11 years) is released into the atmosphere. By the year 2000, the estimated annual release of krypton-85 will be 250 million *curies*.* In addition, large quantities of tritium and other isotopes may be released into aquatic environments.

What are the possible dangers to the environment and to life from radioactivity released from power plants? Even though radiation escaping from nuclear power plants is usually of a very low level, certain isotopes remain in the environment for dangerously long periods of time. The long-term effects of low-level radiation on man's genetic make-up are not well un-

---

*A curie is the amount of material which produces 2.2×10^{12} disintegrations per minute.

derstood. It is known, however, that any amount of radiation above background level usually causes an increased rate of mutations — most of which lead to undesirable characteristics in the offspring of future generations. Furthermore, radioactive materials can be concentrated by organisms in the food web. A study in the Columbia River indicated that radioactivity of the water, mostly from the reactors of the Hanford Nuclear Center, was slight; but radioactivity of river plankton was 2000 times greater than that of the water; of plankton-feeding fish 15,000 times greater; of young insect-feeding fish 15,000 times greater; of young insect-feeding swallows 500,000 times greater; and of waterfowl egg yolks from river-inhabiting birds 1 million times greater than the radioactivity level of the river.

What are the biological consequences of exposure to radiation? Human exposure to *high* doses of radiation can result in severe symptoms including temporary or permanent sterility, cancer, leukemia, acute anemia, cataracts, loss of hair, retarded growth, blindness, epilepsy, brain damage, shortening of life span, severe burns, and death. But *any* exposure, even to low levels of radiation, can increase genetic mutations and affect the offspring of exposed persons. Such genetic defects may include hemophilia, allergies, dwarfism, clubfoot, harelip, Siamese twins, and many other gene mutations or chromosome aberrations (Fig. 7–13). Increasing the dosage of radiation increases the genetic damage.

What are the effects of long-term exposure to low rates of radiation between 0.1 rad per year (natural background) and 5 rads per year (the maximum permissible exposure for radiation workers)? It is known that tumors can be induced in mice by radiation of 50 rads or more, but studies on populations large enough to establish a positive relationship with lower levels of radiation are not numerically feasible. It is known, however, that children born to mothers who had abdominal x-rays during early pregnancy have about a 50 per cent greater probability of dying from leukemia within the first eigh years of life than children whose mothers were not x-rayed, even if the x-ray doses were only a few rads.

The International Commission on Radiological Protection (ICRP) recommends a maximum allowable *genetic dose* of 5 rads for the general population (per 30-year reproductive period). Drs. Lindop and Rotblat of the University of London have estimated that such an "allowable" radiation dosage applied to the population of the United States would result in 2500 additional cases of cancer every year. John Gofman and Arthur Tamplin of the University of California's Lawrence Radiation Laboratory also criticize existing permissible radiation levels as being too high. They calculate that if the entire population were exposed to radiation levels permitted by the Federal Radi-

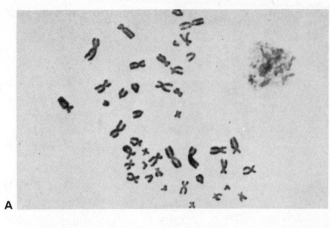

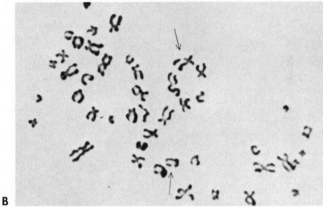

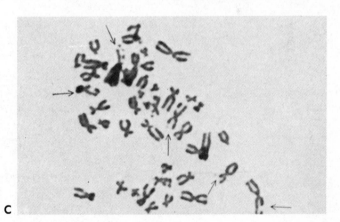

Figure 7-13 Breaks in chromosomes of human tissue cells caused by radiation are marked by arrows in two lower photomicrographs. Chromosomes from an unirradiated control cell (*A*) show no breaks; two breaks are marked in chromosomes from cell irradiated with 50 roentgens (*B*) and five breaks are marked in chromosomes from cell exposed to 75 roentgens (*C*). (From Puck, T.: Radiation and the human cell. Sci. Am. *202*:142–153, April 1960. Reprinted with permission of Theodore C. Puck, University of Colorado. Photomicrographs by the author.)

ation Council, 32,000 additional cases of cancer would result. They argue that there is no justification for accepting *any* additional deaths in the population from radiation pollution.

How Safe Are Nuclear Power Plants?   The United States Atomic Energy Commission (AEC) has spent tens of millions of dollars in the past 25 years in seach of a solution to insure reactor safety. But safety research seems to lag behind actual implementation, while public concern and debate about the safety of nuclear power plants continue. Conflict over possible ill effects of nuclear generated electricity has aligned promoters and advocates of nuclear power (the AEC, some scientists, and the utility companies) on one side and critics (some employees of AEC, various conservation groups, and others) on the other side. However, the general public remains without adequate information to balance the benefits against the risks. The AEC expends large sums of money to educate the public and promote nuclear energy as a safe, practical, economical way to generate electricity.

Many safety devices are built into nuclear reactors to prevent escape of radioactivity into the environment. Proponents of nuclear energy argue that the probability of an accident releasing large amounts of radioactivity is extremely small. Critics, on the other hand, argue that after a point, increasing the number of safety systems on the reactor can only increase its complexity and result in a greater risk of accidents. Nuclear power plants have a good safety record, but a single accident could be catastrophic. A 1-million kilowatt reactor, after operating for one year, contains on the average more radioactive strontium, cesium, and iodine than were released during all the nuclear weapons tests conducted in the world up to the present time. Critics of nuclear energy cite several possible causes of power plant accidents, including mechanical failure, human error, natural disaster, sabotage, and nuclear attack.

Alvin Weinberg, director of the Oak Ridge National Laboratories, has outlined the highly improbable sequence of events that could result in a catastrophic accident at a nuclear reactor. First, a pipe might break or a safety system fail to respond when called upon in an emergency. Second, the core cooling system might fail, the fuel might melt, react with the water, and conceivably melt through the container. Third, the container itself might fail catastrophically or might melt, and overpressurization and explosion could spread radioactivity to the surrounding area.

Several near-miss accidents have occurred. In September 1970, the safety system of the Hanford nuclear reactor failed when called upon, but the back-up safety system operated as planned and shut off the reactor. The accident at the Enrico Fermi reactor near Monroe, Michigan, is often cited as an example of what could happen during a power plant mishap. Plans for the Fermi reactor, an experimental "breeder reactor," began in 1955 (Fig. 7–14). The reactor was located only about 30 miles from Detroit, and an early preconstruction study estimated that under the hypothetical conditions, an accident at the reactor could result in the death of as many as 133,000 persons and the injury of 181,000 others if radioactivity escaped. During court battles concerning the relative

Figure 7-14 Machinery dome inside the containment building of the Enrico Fermi I atomic power plant. Construction was completed in 1963, low power testing completed in 1965. It incorporated a liquid sodium–cooled fast breeder reactor. In 1966, following an accident in which a loose piece of metal obstructed the cooling system, the reactor core partially melted and was rendered inoperable for several years. (From Power Reactor Development Company, Newport, Michigan.)

safety of the reactor, construction began. In January 1966, the reactor was operated above 1 megawatt thermal for the first time. About eight months later, an accident led to a tremendous heat build-up and a partial core melt-down. The reactor had to be shut down by inserting control rods in the core. There were no personnel injuries or exposures to above normally accepted dosages of radiation nor were any fission products released to the environment in excess of specification limits. It took more than a year to determine the cause of the accident. In 1957, a reactor at the Windscale Works in England accidentally released fission products from the smokestacks into the air. Authorities were forced to seize and destroy foodstuffs and milk from an area within a 200-mile radius.

The Los Alamos Scientific Laboratories of the University of California estimate that since the beginning of the atomic energy industry there have been at least 34 occasions when power levels of fissionable materials became uncontrollable because of unplanned or unexpected changes in the reactor. These 34 cases have resulted in eight deaths.

A critical nuclear accident could take place in the space of a minute, or faster than human attendants or mechanical devices could avert it. Edward Teller, a prominent nuclear physicist, stated in a report on reactor safety that "a typical nuclear runaway accident may start and be over in times appreciably less than a second."

Transport of radioactive material is another possible source of accidents. Weinberg estimates that by the year 2000, nuclear power production will require the transport of 7000 to 12,000 spent fuel shipments per year from reactors to chemical plants, with an average of 60 to 100 loaded casks in transit at all times. He estimates that with shipping casks presently being designed to ensure safety during transport, the chances of a serious accident in transit before the year 2000 are between 1 in 3.3 and 1 in 333.

Generation of electricity by nuclear power results in nuclear waste material which must be disposed of or stored for very long periods of time. Disposal is attempted by either concentration and containment or by dilution and dispersion.

Can nuclear wastes be stored in sufficient volume for sufficiently long periods of time to allow decay to reach safe levels? Weinberg estimates that by the year 2000 we shall have to store 2.7×10^{10} curies of radioactive materials in the United States. These wastes will be generating 100,000 kilowatts of heat. Certain isotopes such as plutonium-239, which has a half-life of 24,400 years, will be dangerous for perhaps 200,000 years. Storing the wastes in permanent concrete vaults would require continuous surveillance, essentially forever. Rocketing the wastes out of the biosphere into outer space would be expensive and hazardous. The most promising disposal method under current consideration is burying of wastes in deep underground salt deposits. Some radioactive wastes are stored in a salt mine in Kansas at depths of nearly 1000 feet. However, these salt mines have been penetrated by oil well drilling holes. Fears that water seeping through these holes could lead to solution and dispersion of radioactive wastes have brought the feasibility of this plan under question. These deep mine disposal sites will have to be protected for many generations against man's intervention.

During an average lifetime each person served by nuclear powered electricity would account for a maximum nuclear waste accumulation of less than ½ pint in volume. Drs. Starr and Hammond of the University of California estimate that while this per person/US lifetime energy production would represent a value of $10,000 per person, the cost of nuclear waste storage would be only $10. Thus, the volume of storage is small, and the cost of storage is a small fraction of the value of the power. Using the ancient Egyptian pyramids as an example, Starr and Hammond suggest that large stone structures or underground salt mines could contain the volume of nuclear wastes generated by the United States at its present rate of electric power consumption for more than 5000 years.

How can the dangers of radiation pollution be avoided? Lindop and Rotblat propose that to ensure safety in the near future several immediate steps be taken: (1) reduction of the allowable dose limit for individuals and the genetic dose for the population by an order of magnitude; (2) prohibition of the siting of reactors near large population areas; and (3) discouragement of any experiments with nuclear explosives.

Once committed to a long-term, large-scale reliance on nuclear energy for great quantities of power, man is also committed to future generations of vigilance in protecting the environment from radioactive pollution. In this sense his freedom of options and available choice of certain land areas for use are reduced.

Solid Waste

The growth of goods and services in the United States between 1970 and 1980 is estimated to be equal to the growth during the entire period from the landing of the Pilgrims to 1950. Accompanying this growth is a tremendous increase in solid waste: junk cars, cans, bottles, plastic, fly ash, and paper products. New synthetics, new packaging techniques, planned obsolescence, and the "throwaway society" have led to rapidly growing mountains of solid waste materials. About 50 kg of solid waste is generated daily for every man, woman, and child in the United States, and by 1980 this is expected to increase to 68 kg per day.

Solid waste can be divided into three categories. The first of these, *urban refuse,* includes domestic, commercial, municipal, and industrial waste products. This source generates in the United States alone about 400 million tons per year including 60 billion cans, 36 billion bottles, 58 million tons of paper products, 4 million tons of plastics, more than 1 million abandoned automobiles, mountains of demolition debris, 180 million tires, countless millions of tons of refrigerators, stoves, TV sets, and other appliances. The second category, *mineral waste* from mining and mineral processing operations, results in 1.7 billion tons of mineral waste annually. Each year 11 million tons of ferrous metals and over 1 million tons of nonferrous metals are discarded in United States municipal dumps. Automotive scrappage alone produces more than 10 million tons of ferrous and one-half million tons of nonferrous metals annually. The third category, *agricultural waste,* produces over 2 billion tons per year from farms, slaughterhouses, and animal waste. An average sized steer generates about 10 tons of solid waste yearly.

Disposal of Solid Waste.   Solid wastes are disposed of or dispersed either in the air, water, or land. Methods of disposal include open dumping, open burning, ocean disposal, composting, incineration, sanitary land fills, and deep wells.

Dumping and Land Fill. Open dumps are most often an unsatisfactory, unsightly, and unsanitary method of solid waste disposal. The so-called sanitary land fill, where solid waste is periodically covered with a layer of earth, has the advantage of inexpensive operation (where land is available). It can be used to reclaim land for desirable uses such as parks or golf courses. But such land may later be unsuitable for construction because of settling, and if not prepared properly, has the potential of polluting the ground or surface water. Furthermore, available land for such sites is dwindling. Wetlands and marine estuaries, currently frequent targets for land filling, should be reserved for their important ecological and recreational values. In 1971 coastal cities and towns dumped some 48 million tons of waste into the sea. Environmentalists are now calling for laws to ban unregulated dumping of wastes in oceans and lakes.

Incineration. One of the major disadvantages of incineration is its contribution to air pollution. This is especially true for the many inefficient burners in current use and for open burning. Burners may also be expensive to install, operate, and maintain, and may release toxic compounds into the atmosphere. On the other hand, incineration largely reduces the volume of waste and, in some cases, salvageable metals and other materials can be recovered.

Composting. In composting, solid organic wastes are covered and allowed to decompose by bacterial degradation. This method is gaining popularity on a small scale on household farms in the United States. It has been widely used in Europe as a source of rich organic fertilizer. Large-scale composting of municipal wastes could be used to generate methane gas by bacterial fermentation and prove to be desirable and profitable when one considers both the economic and ecological cost of alternative disposal methods.

Recycling. Recycling of some industrial products, especially metals, is already being carried out. For example, one-half the United States requirement of copper and two-thirds of the requirement for lead and zinc are derived from recycling, and in 1969, 41 million tons, or 50 per cent of the material used for the production of all steel products came from scrap. Nevertheless, the wide dispersal of such scrap forms as tin cans, and the fact that much of urban waste is mixed and difficult to separate has discouraged large-scale recycling. In order to stimulate recycling plans, a number of bills have been introduced into the United States Congress and state legislatures to ban or tax disposable containers and put a tax on autos, refrigerators, and other manufactured goods in order to pay for the cost of their ultimate disposal.

The Office of Solid Waste Management Programs of the
EPA actively investigates methods for better control of solid
wastes. Also, the United States Department of the Interior
Bureau of Mines is experimenting with methods for separating
and recovering metals and other substances from scrap materials. Their experimental plant can process one-half ton of residues per hour and uses magnetic separation, screening, grind-
ing, and shredding processes to separate ferrous and
nonferrous metals, and glass from burned refuse on a continuous basis. Estimates for a 1000 ton per day plant indicate a cost
of about two dollars per ton of residue processed. Each ton
would yield 700 pounds of iron, 40 pounds of nonferrous
metals (including aluminum, copper, lead, tin, zinc, and small
amounts of silver), and a half-ton of glass for a total value of 10
to 12 dollars per ton of residues. The remaining finely pow-
dered ash could be used as agricultural fertilizer supplement.
Recovered glass is used to manufacture bricks and a type of
mineral glass wool. Another experimental process produces
more than two barrels of crude heavy low sulphur oil from each
ton of dry refuse. Organic waste fractions such as worn-out
rubber tires, urban refuse, wood bark, sawdust, plastics, and
discarded battery cases can be chemically distilled to yield liquid
and gaseous hydrocarbons, tar, and valuable chemicals (Fig. 7–
15).

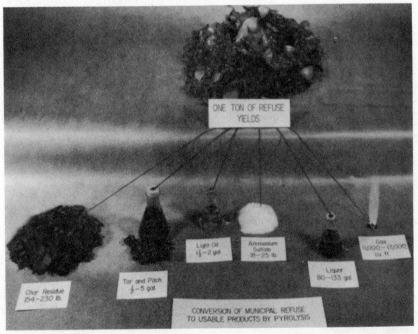

Figure 7–15 Destructive distillation. Organic fractions of municipal wastes can be
pyrolyzed thermally to yield valuable byproducts. (Reprinted with permission from
Kenahan: Solid Waste Environ. Sci. Tech. 7:597, July, 1971. Courtesy of American
Chemical Society.)

THE CAUSES OF POLLUTION

In 1971 Barry Commoner et al. analyzed the question, "What actions of human society have given rise to environmental deterioration?" Paul Ehrlich and others had previously pointed out that pollution can be said to be the result of multiplying three factors: population size, per capita consumption, and an environmental impact index that measures, in part, how wisely we apply the technology that accompanies consumption. Thus we could establish an equation as follows:

$$\text{environmental deterioration} = \text{population size} \times \text{per capita consumption} \times \text{environmental impact per unit of production}$$

Commoner et al. analyzed changes in the levels of specific pollutants, population size, and environmental impact per unit of production in the United States between 1946 and 1968.

First, with regard to pollution during the years 1946 to 1968, significant increases in environmental pollution occurred in several areas. Nitrogen and phosphates discharged into surface waters by sewage increased by 260 and 500 per cent respectively; airborne lead increased by about 400 per cent; deadly nitrogen oxide emissions in Los Angeles County air increased by 530 per cent; the average algae population in Lake Erie increased 220 per cent; and the bacterial count in New York Harbor increased 890 per cent. Commoner estimates that between 1946 and 1968 pollution levels, on the whole, increased roughly between 200 and 1000 per cent. Increases in population between 1946 and 1968 were much lower (43 per cent) than increases in total pollution and were thus not sufficient alone to explain the large increases in pollution. If increase in population is not enough to explain an increase in pollution, how about the second factor in the equation (per capita consumption)? Over the years 1946 to 1968 the per capita gross national product (a rough measure of production and consumption) increased 59 per cent, and this, by itself, is again not great enough to explain the great increase in pollution.

Neither increases in population nor in overall per capita consumption of goods is adequate to account for the large increases in pollution. Therefore, Commoner examined the third factor in the equation; that is, the impact on the environment of new technologies used to produce these goods and services. By analyzing the growth in the production of specific goods, Commoner identified several general classes of industrial production that have shown particularly rapid increases in recent years (Table 7–2). These include the production of syn-

TABLE 7-2   Changes in Production or Consumption Per Capita*

Item	Period	% Increase
Nonreturnable beer bottles	1946–69	3778
Mercury for chlorine and sodium hydroxide products	1946–68	2150
Noncellulosic synthetic fiber (consumption)	1950–68	1792
Plastics	1946–68	1024
Air freight—ton-miles	1950–68	593
Nitrogen fertilizer	1946–68	534
Synthetic organic chemicals	1946–68	495
Chlorine gas	1946–68	410
Aluminum	1946–68	317
Detergents	1952–68	300
Electric power	1946–68	276
Pesticides	1950–68	217
Total horsepower	1950–68	178
Wood pulp	1946–68	152
Motor vehicle registration	1946–68	110
Motor fuel (consumption)	1946–68	100
Cement	1946–68	74
Truck freight—ton-miles	1950–68	74
Total mercury (consumption)	1946–68	70
Cheese (consumption)	1946–68	58
Poultry (consumption)	1946–68	49
Steel	1946–68	39
Total freight—ton-miles	1950–68	28
Total fuel energy (consumption)	1946–68	25
Newspaper advertisements (space)	1950–68	22
Newsprint (consumption)	1950–68	19
Meat (consumption)	1946–68	19
New copper	1946–68	15
Newspaper news (space)	1950–68	10
All fibers (consumption)	1950–68	6
Beer (consumption)	1950–68	4
Fish (consumption)	1946–68	0
Hosiery	1946–68	−1
Returnable pop bottles	1946–69	−4
Calorie (consumption)	1946–68	−4
Protein (consumption)	1946–68	−5
Cellulosic synthetic fiber (consumption)	1950–68	−5
Railroad freight—ton-miles	1950–68	−7
Shoes	1946–68	−15
Egg (consumption)	1946–68	−15
Grain (consumption)	1946–68	−22
Lumber	1946–68	−23
Cotton fiber (consumption)	1950–68	−33
Milk and cream (consumption)	1946–68	−34
Butter (consumption)	1946–68	−47
Railroad horsepower	1950–68	−60
Wool fiber (consumption)	1950–68	−61
Returnable beer bottles	1946–69	−64
Saponifiable fat (for soap products)	1944–64	−71
Work animal horsepower	1950–68	−84

*From Commoner, 1971, p. 7.

thetic organic chemicals and the products made from them; wood, pulp and paper products; the total production of energy (especially electric power); total horsepower of prime movers (especially petroleum driven vehicles); and cement, aluminum, mercury, and petroleum products. Increases in these activities may be responsible for the observed changes in pollution levels. Substitution of synthetic organic products for natural ones through modern chemical technology has considerably intensified environmental pollution. Synthetic plastics, detergents, pesticides, and so forth are often not degraded or only slowly degraded by natural decomposition in the environment.

Commoner et al. conclude that

. . . the predominant factor in our industrial society's increased environmental degradation is neither population nor affluence, but the increasing environmental impact per unit of production due to technological changes.

We must agree with Commoner that quantitatively speaking, the environmental impact per unit of production may be very important, but we should also note that dynamically it is increasing population and per capita consumption that impel technological change and innovation.

POLLUTION AND SOCIETY

To a great extent, pollution results from social values and priorities in a technological age — it is a behavioral problem of a materialistic society. Modern industrial societies could not exist without producing some waste materials and untrapped heat energy. However, technology can provide the means for reducing pollution while at the same time maintaining high levels of per capita agricultural and industrial production. Factories, power plants, and transportation facilities with lower pollution production can be achieved if man is willing to pay the added cost. If he does not begin to pay now, the cost will be transferred to nature and result in degraded ecosystems and a breakdown of essential life-support systems. If he allows this to happen, he will eventually be faced with the ultimate cost — the cost paid by many past species — the cost of extinction.

The burgeoning growth of the types of pollution discussed in this chapter requires increasing regulation and control on both a personal and governmental level (see Chap. 10). Legislation to control pollution must, in some way, lead to controls on the types of materials produced by industry, on the locations of disposal, on the waste products that may be released into the environment, and on the quantities of materials released. If we are to adequately control pollution, certain individual freedoms will have to be curtailed. Current pollution-controlling restrictions on freedom include laws against open trash burning, the production of excessive noise, and dumping waste in certain areas. If generation of pollution continues to grow, controls will have to be expanded to limit freedoms on a larger scale. For example, restrictions against driving automobiles in certain areas or at certain times of day are already being considered or implemented in some communities. In many subtle ways, growing pollution challenges individual freedom.

REFERENCES

Beranek, L. L.: Noise. Sci. Am. *215*(6):66–76, 1966. (A general review of the sources, effects, and control of noise pollution.)
Berland, T.: The Fight for Quiet. Englewood Cliffs, New Jersey, Prentice-Hall, Inc., 1970. (General problems of noise pollution.)
Brook, A. J., and A. L. Baker: Chlorination at power plants: impact on phytoplankton productivity. Science *176*:1414–1415, 1972. (Chlorination significantly reduces primary productivity of phytoplankton.)
†Brunner, J.: The Sheep Look Up. New York, Ballantine Books, Inc., 1972 (paperback). (Fictional future world of pollution and environmental degradation.)
Bryerton, G.: Nuclear Dilemma. New York, Ballatine Books, Inc., 1970 (paperback). (The pros and cons of nuclear power generation. Dangers as well as need for nuclear generated electrical power; nuclear and thermal pollution.)
*Clark, J. R.: Thermal pollution and aquatic life. Sci. Am. *220*(3):19–27, 1969. (Review of the problem, particularly effects on fish.)
Cole, L. C.: Thermal pollution. Bioscience *19*(11):989–992, 1969. (Overall possible long-term effects on biosphere and climate of heat released by man.)
Committee on Thermal Pollution: Bibliography on thermal pollution. J. Sanit. Eng. Div. Amer. Soc. Civ. Eng. *9*(SA 3):35–113, 1967. (Extensive list of articles dealing with thermal pollution.)
Commoner, B., M. Corr, and P. J. Stamler: The causes of pollution. Environment *13*(3):2–19, 1971. (Pollution results not primarily from population growth, or affluence, but from new production technologies and substitution of synthetic products.)
Craig, P. P., and E. Berlin: The air of poverty. Environment *13*(5):2, 1971. (Distribution, sources, and effects of lead pollution.)
Ehrlich, P. R., and A. H. Ehrlich: Population, Resources, Environment: Issues in Human Ecology. San Francisco, W. H. Freeman and Co., 1970. (World human population, growth, resources, pollution, etc.)
Hall, S.: Pollution and poisoning. Environ. Sci. Tech. *6*(1):31–34, 1972. (High lead levels are dangerous to man, and concentrations are rising.)
Hammond, A.: Fission: the pro's and con's of nuclear power. Science *178*:147–149, 1972.
Kenahan, G. B.: Solid waste—resources out of place. Environ. Sci. Tech. *5*(7): 594–600, 1971. (The US Bureau of Mines experiment with recovering solid waste materials.)
League of Women Voters: Solid waste—it won't go away. Current Focus, Publ. No. 675, 1971 (available 50c. 1730 M St., NW, Washington, DC 20036). (Review of the problem, including a good list of further information.)
*Lindop, P. J., and J. Rotblat: Radiation pollution of the environment. Bulletin of Atomic Scientists, September 1971, pp. 17–24. (Concentrates on the hazards to man of radiation exposure.)
McCaull, J.: Building a shorter life. Environment *13*(7):3, 1971. (Sources, extent, and health effects of cadmium pollution.)
*Odum, E. P.: Fundamentals of Ecology. Philadelphia, W. B. Saunders Co., 1971. (Textbook of general ecology.)
Patrick, R.: Some effects of temperature on freshwater algae. In Krenkel, P. A., and F.

*Recommended further reading.
†Fiction

L. Parker: Biological Aspects of Thermal Pollution. Nashville, Tennessee, Vanderbilt University Press, 1969. (Heated waste-water from power plants causes change in dominance of algal vegetation.)

Puck, T. T.: Radiation and the human cell. Sci. Am. *202*:142–153, April 1960. (Human cells in culture are very sensitive to radiation dosage.)

Subcommittee on Air and Water Pollution of the Committee on Public Works: Thermal pollution. Washington, DC, US Government Printing Office, 1968. (US Senate subcommittee assesses the scope of the problem.)

Sylvester, J. R.: Possible effects of thermal effluents on fish: a review. Environ. Pollut. *3*(3):205–215, July 1972. (Review of the scientific literature.)

Teal, J. M.: Energy flow in the salt marsh ecosystem of Georgia. Ecology *43*(4):614–624, 1962. (Research report of study of food web and energetics.)

Trembley, F. J.: Effects of cooling water from steam electric power plants on stream biota. In US Department of Health, Education and Welfare: Biological Problems in Water Pollution. 999WP-25, pp. 334–335. Washington, DC, US Government Printing Office, 1965. (Heated waste-water reduces species diversity of algae.)

*Turk, A., J. Turk, and J. T. Wittes; Ecology, Pollution, Environment. Philadelphia, W. B. Saunders Co., 1972. (Excellent review of various types of environmental pollution.)

Turk, A., et al.: Environmental Science. Philadelphia, W. B. Saunders Co., 1974. (Excellent well-illustrated introductory text on general ecology, pollution, and the human environment.)

*US Environmental Protection Agency Office of Noise Abatement and Control: The social impact of noise. Washington, DC, US Government Printing Office, December 31, 1971. (Effects of noise on society.)

Weinberg, A. M.: Social institutions and nuclear energy. Science *177*:27–34, 1972. (Nuclear power generation poses problems for man and his environment, but they can be solved.)

Whittaker, R. H.: Communities and Ecosystems. New York, Macmillan Publishing Co., Inc., 1970. (General ecology text.)

Woodwell, G. M.: The ecological effects of radiation. Sci. Am. June 1963. (Ecosystems, particularly forests, are destroyed by ionizing radiation.)

Woodwell, G. M.: The energy cycle of the biosphere. In Scientific American Magazine Editors, eds.: The Biosphere. San Francisco, W. H. Freeman and Co., 1970. (General description of ecosystems and energy flow, from articles in Scientific American.)

Zero Population Growth: A Teacher's Guide to Materials on Population. Denver, Colorado, ZPG, PO Box 18291, 1971. (An excellent source of further materials on population and resources. Includes bibliography of books, audio-visual materials, and population games.)

---

*Recommended further reading

PROBLEMS

1. Why do nondegradable materials released by man into the environment often become more concentrated in the tissues of higher trophic (feeding) level animals, especially carnivores?

2. Why are simplified ecosystems with a reduced number of species more subject to pestilence, disease, or large fluctuations in the numbers of individuals than complex ecosystems?

3. Why does reusing products such as beverage bottles have less impact on the environment than recycling?

4. Design a proposed scheme that you believe would be equitable and effective in controlling a wide variety of types of

pollution in your community. Submit the scheme to the class for discussion of its advantages and disadvantages.

5. Can environmental pollution be totally eliminated by using advanced technology?

6. Has industrialization (including resultant pollution) caused a net increase or decrease in public health? How has this happened?

7. What are some of the "costs" of environmental pollution to a human community? How could these costs be measured for specific types of pollution?

8. List specific items used or activities engaged in by you today (or yesterday) which resulted in some of the types of pollution described in this chapter. Can you think of ways these could have been avoided?

9. Is there such a thing as nonpolluting transportation or energy production?

10. If synthetic products have a greater environmental impact per unit, can we not return to using more natural, biodegradable products?

11. It is said that pollution results from man's desire for convenience. Explain and give some examples.

BRAINSTORM

It is the year 2000. You remember older people telling you how the new green leaves suddenly sprang from trees after winter and new flowers came into bloom. The morning news tells you that the government has announced that due to air and water pollution 75 per cent of trees will not produce buds this year. The Department of Fisheries reports that oceanic plankton are dying due to insecticide and herbicide poisoning, massive oil spills, and high-temperature waste-water from industries. The Department of Health has announced the necessity of using government-issued SCOBA (self-contained oxygen breathing apparatus) within the city because of decreasing oxygen levels and increasing air pollution.

(Adapted from ZPG, 1971.)

8

POPULATION

*In ancient times, people were few but wealthy and without strife.
People at present think that five sons are not too many, and each son
has five sons also and before the death of the grandfather there are
already 25 descendants. Therefore people are more and wealth is less;
they work hard and receive little. The life of a nation depends upon
having enough food, not upon the number of people.*

Han Fei-Tzu
Chou Dynasty, ca. 500 BC

*. . . the only way we can preserve and nurture other and more
precious freedoms is by relinquishing the freedom to breed and
that very soon.*

Garrett Hardin, 1968 AD

POPULATION GROWTH

How Did the Human Population Reach Its Present Level?

It is an ironic fact that some of the most important problems facing mankind today may not be amenable to scientific or technological solutions—one such problem is the population crisis. The rapid growth of the human population has long been recognized as a potential problem. In 1798, Thomas Malthus observed that the means of subsistence (resources, food, and so forth) grow in an arithmetic fashion; that is, their growth is additive. Thus, a farmer with two acres of land under cultivation can add two acres, increasing food production to four acres, then by adding two more acres, to six acres. This sort of additive progression results in straight line growth (Fig. 8–2A). In contrast, populations of organisms, including man, if unchecked, grow not arithmetically but geometrically. In geometric growth, whether it is growth of compound interest on money or the growth of the human population, numbers are not additive but multiplicative. Thus, if two people have two children and both of these children in turn produce two

Figure 8–1 "Excuse me, sir, I am prepared to make you a rather attractive offer for your square." (Drawing by Weber; © 1971 The New Yorker Magazine, Inc.)

children we must multiply $2 \times 2 \times 2$ to determine the final number of people (eight). When factors do not limit this geometric increase of organisms (people) a rapidly accelerating growth curve occurs (Fig. 8–2*B*).

Malthus illustrated just how much greater the power of unchecked geometric over arithmetic growth can be:

... the human species would increase as the numbers, 1, 2, 4, 8, 16, 32, 64, 128, 256, and subsistence as 1, 2, 3, 4, 5, 6, 7, 8, 9. In two centuries the population would be to the means of subsistence as 256 to 9; in three centuries as 4096 to 13, and in two thousand years the difference would be almost incalculable.

Such rapid geometric growth, whether in a population of fruit flies or of humans, cannot continue indefinitely. Growth is

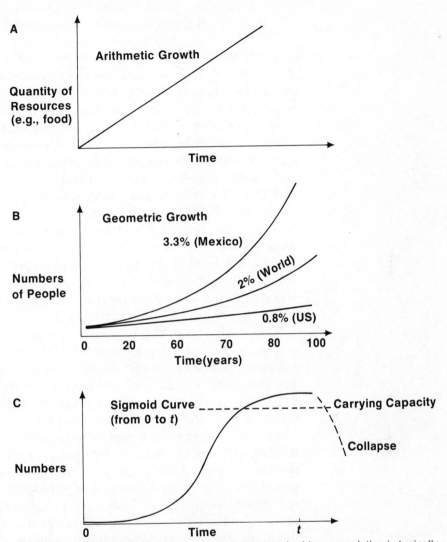

Figure 8–2 *A,* Growth in the supply of resources required by a population is typically arithmetical. Additional acres of food under cultivation, new oil wells, new power-generating plants, can be added one by one.

    B, In biological organisms, an abundance of required resources (e.g., energy, space, nutrients) typically leads to a high reproductive and survival rate. The positive feedback (more surviving offspring ⟶ more adults of reproductive age) results in *geometric* (or exponential) *growth* where numbers grow much faster than in *A.*

    C, All resources are finite. As a required resource (or resources) is depleted by a growing population, the scarcity of the resource slows down the rapid geometric growth of curve *B* (by acting as a negative feedback loop). Finally, growth becomes negligible as the population reaches the "carrying capacity" of the environment. At that point, the limiting resources is not available in sufficient quantities to allow further growth. In some cases rapid geometric growth can lead to an oscillation above and below the carrying capacity for some time.

limited by the availability of essential resources. Eventually, factors in the environment such as crowding, disease, depletion of food resources, or a combination of such factors, halts further population growth, resulting in a sigmoid growth curve (Fig. 8–2C). The growth rate or *rate of natural increase* of a population is simply the *birth rate minus the death rate* (discounting immigration to and emigration from another area). Environmental factors may lead to a temporary halt or decrease in population growth, which can then recover and continue to grow. For example, bubonic plague (black death), which periodically invaded Europe between the fourteenth and seventeenth centuries, increased the death rate tremendously and killed an estimated 25 per cent of all Europeans between 1348 and 1350. In the long run, however, the plague only temporarily slowed population growth.

The most remarkable feature in the history of human population growth is the very slow rate of increase that existed for at least the first one to two million years. Studies of archaeological remains and of densities of present day hunting and gathering tribes indicate that during most of man's first million years his numbers probably did not greatly exceed 5 million. Only within the past 500 years did this figure begin to accelerate toward the present level of 3.8 billion (Fig. 8–3A). Estimates indicate that two-thirds of all people born since 1500 are alive today. Until about 1700 the human population grew at an average rate of probably less than .002 per cent per year; the current world average rate (2 per cent) is 1000 times greater. At present the human population is increasing by more than 200,000 people each day or 75 million per year. Such a rate of increase cannot be sustained for more than a moment of geological time. If current growth continues for only a few thousand more years, everything in the visible universe will be filled with people, and this mass of people will be expanding with the speed of light. The recentness of this growth becomes evident when we consider that nearly 3 per cent of all the people that have ever lived are alive today. If the present rate of growth had existed since the time of Christ, the world population would have increased by now to 7×10^{16} or about 100 people for each square foot of the earth. In short, current population growth cannot continue.

Historically man's expansion probably occurred in three major steps. More than one million years ago our early relatives numbered only about 125,000 and were confined to the continent of Africa. The species was not much more prevalent than any other large mammal of the period. However, the evolutionary development of both culture and increasing brain size in man (see Chap. 4) soon led to a rapid increase

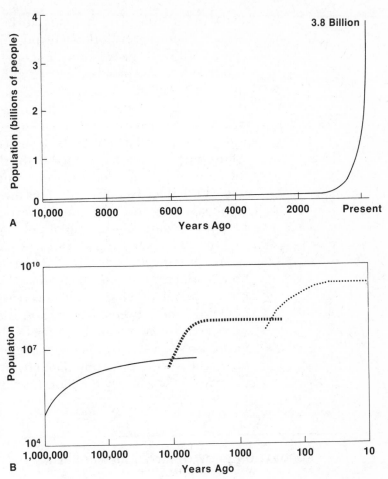

Figure 8–3 *A,* Arithmetic population curve plots the growth of human population from 10,000 years ago to the present. Such a curve suggests that the population figure remained close to the base line for an indefinite period from the remote past to about 500 years ago, and that it has surged abruptly during the last 500 years as a result of the scientific-industrial-medical revolution. *B,* logarithmic population curve makes it possible to plot, in a small space, the growth of population over a longer period of time and over a wider range (from 10⁴ or 10,000, to 10¹⁰ or 10 billion persons). The curve, based on assumptions concerning the relationship of technology and population, reveals three population surges, reflecting toolmaking or the cultural revolution (*solid line*), the agricultural revolution (*broken line*), and the scientific-industrial-medical revolution (*dotted line*).

in population. The development of tools, increased intellectual capacity, and language allowed man to transfer knowledge from generation to generation, and probably provided the impetus for the initial growth in population that began about one or two million years ago. Man's migrations led him to occupy all the major continents of the earth; the population reached approximately five million and remained near this level until about 8000 BC. At this time the second major development in man's evolutionary history occurred — the first agricultural revolution, which began around 9000 to 7000 BC in the Middle East (and perhaps as long ago as 13,000 BC in Southeast Asia). Man the hunter and food gatherer now turned to cultivating the land and growing his own food. Survival became less difficult, and the human population again took a great leap forward — multiplying an estimated 16 times within 10,000 years. About 300 years ago (only an instant on the evolutionary time scale), a third stimulus to the population growth occurred. This was the scientific-in-dustrial-medical revolution. During this revolution (which spread outward from western Europe) average life expectancy at birth increased progressively from perhaps 25 to 30 years until it reached about 41 years in the first half of the nineteenth century. This was principally due to a rapid decrease in the death rate caused by improved living conditions, while the birth rate remained relatively stable or showed only a slight decrease. Thus, the population growth rate (birth rate minus death rate) increased drastically, leading to a fivefold increase in population within the past 300 years. This three-step increase in population resulting from three major revolutions becomes more evident in a logarithmic graph of population growth (Fig. 8–3*B*).

Instead of speaking of the growth rate of populations, we can refer to the *doubling time* or time required for the population to double in size (see box insert). The doubling time of the human population has steadily decreased from an average doubling time between 8000 BC and 1650 AD of about 1500 years to the present doubling time of about 35 years. This means that 35 years from now (if present growth rates continue) the world will be populated by twice as many people as there are now—nearly 8 billion human beings.

POPULATION GROWTH RATES AND DOUBLING TIME

The population growth rate indicates the speed of population growth. In describing human populations, it is usually expressed as *per cent per year*. It is analogous to bank interest rates, which indicate the speed at which deposited money grows.

Population growth rates are determined by the difference between the birth (natality) rate and the death (mortality) rate (excluding immigration and emigration), which are usually expressed as so many births or deaths per thousand population (0/00) per year. The population growth rate in per cent per year may be calculated by subtracting the death rate from the birth rate and dividing by 10.

Example: Birth rate = 22 babies born per thousand people
per year = (22 0/00)

Death rate = 10 people die per thousand people
per year = 10 0/00)

Growth rate = $\dfrac{22 - 10}{10}$ = 1.2 per cent per year

Doubling time is the amount of time needed for a population to double (e.g., to go from 2 to 4 or from 3.7×10^3 to 7.4×10^3).

The following are examples of doubling times:

	Population Growth Rate (%)	Doubling Time (years)
United States	0.8	87
World	2.0	35
Costa Rica	3.8	19

The approximate time required for a human population to double, assuming constant natality and mortality, can be calculated by dividing 70 years by the population growth rate in per cent.

Example: World population growth rate $= 2\%$
World population doubling time $= 70$ years $\div\ 2 =$
35 years.

Population growth rates as well as affluence differ greatly between regions of the world. The current growth rate of the United States population is 0.8 per cent; the average for the world is about 2 per cent; and for many underdeveloped countries (UDCs) the average is greater than 3 per cent. In general, the population growth rate for the so-called UDCs or poor countries is much greater than that of the richer industrially developed countries (DCs).

Let us first examine recent population growth patterns in the industrialized (developed) countries. Although occurring at different times and at different rates, the modern industrial countries of North America, Europe, the USSR, and Japan followed a similar three-stage population growth pattern. The first stage of this pattern was characterized by high birth rates and high death rates. In medieval Europe, for example, estimates indicate an average life expectancy of about 27 years. Death in childbirth, disease, starvation, and wars all contributed to keep the death rate high and the natural rate of increase (despite high birth rates) low. During the second stage of this pattern, called a *demographic transition*, gradual industrialization and improvements in medicine, public health, and agriculture led to steady increases in the average life expectancy. With a gradual decrease in the death rate, populations grew more rapidly, and between 1650 and 1750 it is estimated that the population of Europe and Russia increased from 103 million to 144 million. Soon, however, the growing industrialized countries entered the third stage of population progression, characterized by a decline in the birth rate. The decline in birth rate in some areas eventually equaled the earlier decline in the death rate, resulting once more in a very slow rate of population growth. This decline in the birth rate resulted from primitive birth control measures, postponement of marriage, and even celibacy. It most probably did not result from poverty or hunger, since in most cases economic prosperity and affluence grew faster than the population.

Kingsley Davis of the University of California believes that the main force behind this decline was "the desire of the people involved to preserve or improve their social standing by grasping the opportunities offered by the newly emerging industrial society." In agricultural societies children are considered economic assets — they contribute to farm labor and provide for their aging parents. In industrialized societies, however, children become an economic deficit — they simply have to be provided for and educated. Thus, the modern industrial nations reduced their birth rate and offset to some extent the decrease in the death rate brought about by modern medicine and public health measures. In Sweden the present population growth rate is about 0.3 per cent and in West Germany 0 per cent (replacement level). These large-scale demographic transitions in Europe and elsewhere began *before* the development of truly modern birth control methods.

The recent history of population growth in the underdeveloped countries is in marked contrast to the pattern of the modern industrial countries. Since 1930 the rate of population growth in underdeveloped countries has been about twice that of the developed countries. In fact, the present population growth rate in UDCs is much greater than ever experienced in the history of DCs. Whereas reduction of death rates through modern medicine took years to develop in the industrialized countries, it has recently been exported rapidly to the underdeveloped countries. The introduction of modern drugs to control diseases such as malaria, yellow fever, smallpox, and cholera has resulted in accelerated decreases in the death rate in UDCs. For example, on the island of Mauritius, in the Indian Ocean, the average life expectancy was increased from 33 to 51 years within an eight-year period following World War II. This same increase took Sweden at an earlier time 130 years to achieve. The rapid decrease in death rate in UDCs has not, however, been accompanied by a corresponding decrease in the birth rate. The result has been extremely high rates of population growth (Fig. 8–4). Furthermore, it is doubtful that the underdeveloped countries can soon achieve the drastic decrease in birth rate necessary to stabilize their population growth. In rapidly growing populations such as those of the UDCs, high birth rates signify large numbers of people in the lower age brackets, resulting in a low average age for the population and a high *dependency ratio.* This means that large

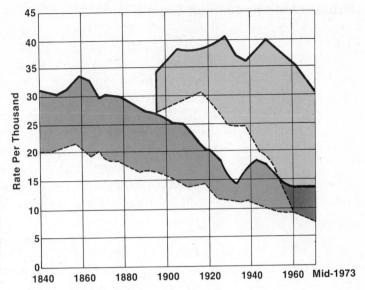

Figure 8-4 In economically under-developed countries such as Sri Lanka (Ceylon), the formerly high birth rate (*solid line*) has decreased only slightly, while the rapid introduction of modern medicine and public health improvements after 1920 has led to a rapid decrease in the death rate (*dotted line*), a greater number of survivors, and a rapid population growth rate.

In developed (industrialized) countries such as Sweden, the birth rate has decreased steadily over a period of more than a century. The death rate decreased more slowly as modern medicine was gradually developed and applied. The present narrow gap between the birth rate and death rate represents a low population growth rate.

numbers of people are within the non-working-age groups of 0 to 14 or 65 and over; people who for the most part are supported by others. In Costa Rica, about 50 per cent of the population falls within these dependent age groups, whereas in Japan, only 31 per cent are in this group (Fig. 8-5). Such high dependency loads that result in high proportions of nonproductive people in the population may inhibit the development of some of the primary incentives for birth control—education and affluence.

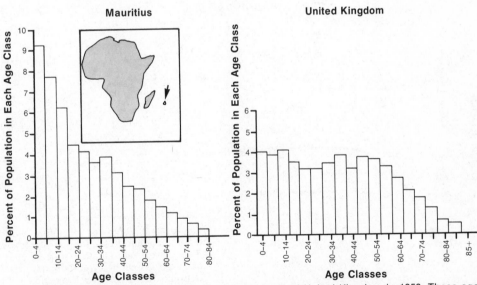

Figure 8-5 Age structure of population of Mauritius and United Kingdom in 1959. These age profiles contrast the age distribution in a rapidly growing underdeveloped country (UDC) with a very slowly growing developing country (DC). In Mauritius young people predominate; in the United Kingdom the population is more evenly distributed over the age spectrum. Note that in each profile the percentage of males of each age class in the population is shown to the left of the center line and that of females to the right.

Future World Population Growth

Data gathered by the United Nations provide low, medium, and high projections of future population growth, all based on the assumption that birth rates will decline in those areas where they are now extremely high. According to the mid-1973 United Nations *medium* forecast, the world's population will increase from its present 3.8 billion to about 6.5 billion by the year 2000. United Nations "constant fertility" projections, which assume that current trends in birth rate and death rate will continue, indicate an even larger growth of the world population — to a level of about 8 billion by the turn of the century.

Tomas Frejka has provided world population predictions based on five different assumptions about when a net reproduction rate of 1 will be achieved and maintained. A net reproduction rate of 1 indicates that the fertility and mortality conditions are such that the current generation of childbearers will be replaced by a generation of the same size. Fertility corresponding to this value is called *replacement fertility* and in developed countries equals about 2.1 children per couple.

The five possible projections differ widely (Fig. 8–6). However, the minimum (projection 1) based on worldwide achievement of replacement level fertility between 1970 and 1975 can

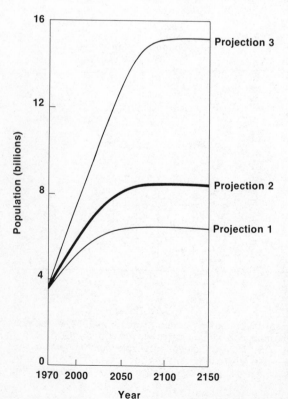

Figure 8–6 Estimates of future world population differ according to the time at which a world replacement level reproduction rate is assumed to be achieved. If such a reproduction rate is achieved between 1980 and 1985, Projection 1 is likely. If such a rate were not attained until 2040 to 2045, Projection 3 would occur. However, the most probable projection is that a replacement reproduction rate will be reached between the years 2000 and 2005, and the population will stabilize at more than 8 billion. (Adapted from Frejka, T.: The prospects for a stationary world population. Sci. Am. *228*(3):15–23, 1973.)

be ruled out as unlikely. At the other extreme, if a net reproduction rate of 1 were attained only very gradually by the middle of the next century, a population of 15.1 billion would be reached — this also seems improbable. The more probable trend, achievement of a net reproduction rate of 1 between 2000 and 2005, would result in the leveling off of the world population at nearly 8.4 billion by the year 2100. In any case, Frejka believes it is either highly unlikely or impossible that zero population growth will be achieved for the world by the year 2000; to reach this goal, the total fertility rate throughout the world would have to drop far below replacement level.

The consequence of markedly different growth rates in the two segments of the world's population (the UDCs and DCs) will result in a change in their size ratio. The major areas of southern Asia, Africa, and Latin America have a high potential for further population growth. Thus, the present ratio of 30:70 between the populations of the rich and poor countries will swing inexorably to 20:80 or even perhaps to 10:90.

Past population growth predictions have tended to err on the low side. Nevertheless, some demographers (scientists who study population) hold the opposite view; and population biologist Paul Ehrlich says,

> We feel that 1963 U.N. projections for the year 2000, with the possible exception of the lowest forecast, are too high. This is not, however, because we share their optimism about the impact of family planning programs on birth rates. Instead, we predict that a drastic rise in the death rate will either slow or terminate the population explosion.

Resources

Can We Provide a Reasonable Quality of Life for the Growing World Population?   The present average world population growth rate is about 2 per cent with a doubling time of approximately 35 years, but as we have seen, this growth differs considerably between regions of the world. The developed industrial countries are growing much more slowly than the nonindustrial, undeveloped countries, and the economic advancements in the undeveloped countries are not keeping pace with their population growth. Thus, in the future, we can expect to find an ever increasing proportion of the world's population in the underdeveloped, poor countries. In 1930 the DCs constituted 32.7 per cent of the people in the world; in 1965 the figure had dropped to 27.5 per cent; and (according to United Nations constant fertility projections) by the year 2000 DCs may contain only 18.8 per cent of the world's population. Unless the re-

TABLE 8–1   World and Regional Population (millions)*

	mid-1973	% of Total	Year 2000 UN Medium Estimate	% of Total	% Increase 1973–2000	Change in % of Total 1973–2000
World	3860	100.0	6494			
Asia	2204	57.1	3777	58.1	71.4	+1.0
Europe	472	12.2	568	8.7	20.3	−3.5
USSR	235	6.5	330	5.1	32.0	−1.0
Africa	374	9.7	818	12.6	118.7	+2.9
North America	233	6.0	333	5.1	42.9	−0.9
Latin America	308	8.0	652	10.0	111.7	+2.0
Oceania	21	0.5	35	0.5	66.7	0.0

*From United Nations Data, 1973, Population Reference Bureau, Inc.

sources (food and energy*) available to the UDCs can be increased rapidly, or unless the resources controlled by the DCs are distributed to the UDCs (both solutions seem highly unlikely), the people of the UDCs will become even poorer as their populations grow and their subsequent per capita share of resources dwindles (Table 8–1).

The world's resources are finite. Our great nonrenewable energy reserves of fossil fuels (coal, oil, and natural gas) and uranium are being depleted rapidly (see Chap. 9). Practically all good agricultural land is already being cultivated by the expanding world population, and soil in many areas, now depleted of its natural nutrients, must be continually supplied with artificial fertilizers to maintain productivity.

The developed countries consume a disproportionate share of these finite resources. The United States, for example, with only 5.4 per cent of the world's population consumes about 35 per cent of world resources. With the current low rate of economic growth in many of the UDCs it becomes doubtful whether they can ever industrialize before many of the world's resources are depleted by the DCs. Thus, as Paul Ehrlich has said, it may be more accurate to refer to underdeveloped countries as "never-to-be-developed countries."

US Population Growth: Pattern and Outlook

In 1972, the President's Commission on Population Growth and the American Future had the following to say about population growth in the United States:

After three years of concentrated effort, we have concluded that, in the long run, no substantial benefits will result from further growth of the nation's population.

*Some developing countries of the Middle East have ample energy supplies of petroleum to last for many decades (see Chap. 9).

Rather that the gradual stabilization of our population
would contribute significantly to the nation's ability to solve
its problems. We have looked for, and have not found, any
convincing economic argument for continued population
growth. The health of our country does not depend on it,
nor does the vitality of business nor the welfare of the
average person.

The commission pointed out that long-term zero population growth (ZPG) must inevitably be established. The question
should only be when and how this goal will be achieved. However, the means used to limit United States population growth
must be consistent with the fundamental values of life; respect
for human freedom, dignity, and individual fulfillment, and
concern for social justice and welfare.

Between 1900 and 1972 the United States population
increased from 76 million to 209 million (1.7 per cent), and
during the same period, the average life expectancy increased
from 47 to 70 years. The present United States population
growth rate is about 0.8 per cent, but with such a large population this rate of growth equals about 1.68 million people each
year—about enough to fill a city the size of Philadelphia.

American women now in their late thirties have, on the
average, more than three children, but there are optimistic
signs of a decrease in this rate. Today's young people expect to
have fewer children: according to a 1971 census bureau survey,
young married women (ages 18 to 24) say that they expect to
have on an average only 2.4 children. There is no reason, however, for complacency regarding United States population
growth. Because of the earlier higher birth rate, relatively large
numbers of young people are still entering the childbearing age
group. Even though they themselves have few children, the
absolute number of people in the population will continue to
grow. Thus, even if immigration and emigration ceased and
couples had only 2.2 children on the average (just enough to
replace themselves and allow for infant mortality), the population of the United States would continue to grow for about 70
years (Fig. 8–7). Potential population growth is often compared
on the basis of either a two- or three-child family projection. If

Figure 8–7 The population of the United States passed the 100-
million mark in 1915 and reached 200 million in 1968. If families
average two children in the future, growth rates will decrease and
the population will reach 300 million in the year 2020. At the 3-child
rate, the population would reach 300 million in this century and
400 million in the year 2020. If the 3-child rate were continued until
2070, it would result in a US population of about 1 billion. (Projections assume small future reductions in mortality and that future
immigration will remain at present levels.) (From Population and
the American Future, 1972.)

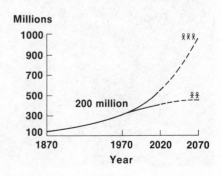

United States families had only two children on the average and emigration continued at current levels, the population would grow to 271 million by the end of the century, but if families were to have an average of three children, the population would reach 322 million by the year 2000. One hundred years from now, the two-child family would result in a population of about 350 million; the three-child family would produce a total of nearly 1 billion. Thus, a difference of only one extra child per family, if extended over a century, would result in an additional 650 million people (Fig. 8–7).

United States population has also undergone a major redistribution. When people who are now 50 years old were born, half the population was rural, but during their lifetime major redistributions in population have occurred. Today, 69 per cent of the American people live in metropolitan areas (cities of 50,000 or more including the surrounding economically integrated county or counties). Thus, in recent decades in the United States as well as elsewhere, a major segment of the population and future reproductive potential has shifted from town and countryside to urban centers (Fig. 8–8).

Figure 8–8 (From Science *169*, Sept. 18, 1970. Photo courtesy of William Garnett.)

The land area occupied by metropolitan centers in the United States has expanded even faster than the number of people occupying these areas. Central cities of many metropolitan areas are actually decreasing in population while suburbs continue to grow rapidly. This dispersal of the affluent (primarily white) segment of population to suburban areas has been accompanied by a concentration of blacks in the central city, so that by 1970, 41 per cent of metropolitan whites and 78 per cent of metropolitan blacks lived in the central cities while the suburbs continued to be almost totally white. This situation—blacks and the poor in the inner city, whites and the wealthy in the suburbs—presents problems of racial and economic separation. The inner cities, faced with limited and sometimes shrinking tax bases because of the urban exodus of affluent whites, are still responsible for the needs of the remaining population—the elderly, the poor, the unemployed, and the nonwhite.

Many Americans are expressing dissatisfaction with city life and believe that the situation should be improved. In a national public opinion survey conducted for the President's Commission on Population Growth and the American Future, over half the respondents felt that the federal government should "discourage further growth of large metropolitan areas" or "should try to encourage people in industry to move to smaller cities and towns." The survey concluded that "Americans are urban and becoming more so, but many people apparently dislike the trend."

The urbanization of population is expected to continue. In 1970, about 71 per cent of the United States population was metropolitan, but by the year 2000 the figure is expected to reach 85 per cent. Population distribution in the United States has proceeded from farm, to small town, to city, to large metropolitan area (Fig. 8–9). It is now becoming concentrated in urban areas of 1 million people or more in continuous metropolitan zones and intervening counties where one is never far from a city. These sprawling semi-cities result not only from increasing population, but also from increasing communication and transportation technology, which encourages the outward movement of industries and residences from the city proper. By the year 2000 supercity megalopolises will occupy about one-sixth of the continental land area of the United States and contain five-sixths of the people. If the commission's projections are correct, "Fifty-four per cent of all Americans will be living in the two largest urban areas. The metropolitan belt stretching along the Atlantic seaboard and westward past Chicago would contain 41 per cent of our total population, while another 13 per cent would be in the California region between San Francisco and San Diego." City planners are now propos-

Figure 8-9 The hexahedron, two mated pyramids over 3500 feet that would house an entire city of 170,000 people on less than half a square mile of land. [Design by Paolo Soleri; photograph by Jon Eaton, Washington, D.C.] (From *Science,* November 28, 1969.)

ing radical new designs to house the crowded urban populations of the near future (Fig. 8–9).

Is Population Growth Necessary for Economic Prosperity? The Commission on Population Growth and the American Future examined this question, and predicted that by the year 2000 per capita income would be as much as 15 per cent higher under the two-child than under the three-child family projection. This results largely from the fact that under conditions of slower population growth, people of working age constitute a greater fraction of the total population. Lower population growth would actually cause total as well as per capita income to be higher over the next 10 to 15 years. However, the commission found that although the general increase in average income associated with slower population growth could assist in reducing poverty, it would not eliminate it. According to the commission, a slower growth rate would not cause problems for any industry or its employees. The diminished burden of providing for dependents and for facilities to keep up with expanding population would make available more of the national output for many desirable purposes, and there would be ampler per capita resources to deal with high priority problems and situations. With respect to the economy, the commission concluded that ". . . in fact the average person will be markedly better off in terms of traditional economic values if population growth follows the two child projection rather than the three child one."

Population growth places demands on key governmental services, causing for example, decreased individual citizen representation and participation in the legislative decision-making process. Each citizen's representation at the federal level is diluted by population growth. In 1910, with a House of Representatives of 435 members, each congressman represented 211,000 citizens on the average; in 1970, each congressman represented an average of 470,000 citizens. By the year 2000 each congressman in a 435-seat House will represent 623,000 persons under the two-child growth rate or 741,000 persons under the three-child growth rate.

Reproductive rates are closely correlated with income and educational levels. In general, birth rates are highest among the poor, lowest among the middle class, and higher again among those in the highest socioeconomic class. Among these highly affluent families, the addition of a third or fourth child poses little economic burden. In this manner, affluent families place greater stress on the environment by consuming a greater quantity of resources and generating larger amounts of pollution per capita than poor families.

If the poor and disadvantaged groups were provided with the same economic, educational, and social opportunities now enjoyed by the middle classes, their currently high birth rates might be expected to decrease. Although the average birth rates of minority groups and the poor are higher than that of middle-class whites, population problems cannot be solved by simply inducing the "have-not" groups to limit their number of children. Because of their smaller numbers, minorities, despite higher fertility rates, contribute less to population growth than the rest of society. For example, even if no babies had been born to black or Spanish-speaking parents throughout the 1960s, the United States population would be only 4 per cent smaller than it is today. As the commission states, "The idea that our population growth is primarily fueled by the poor and minorities having lots of babies is a myth."

Years of segregation and exclusion have led to feelings of powerlessness and alienation among many minorities. They believe that overpopulation is a predicament of white affluent Americans and is irrelevant to the survival problems faced by blacks and other minority groups. People from minorities who have "made it" in the system tend to adopt a small family pattern. The task, then, is to make the system work for members of minority groups as it has for the middle class. We must address our major domestic and social problems beginning with racism and poverty in order to fully resolve the dilemma of overpopulation.

The benefits and rewards of raising children are well known, but not many people recognize the emotional and financial costs involved. Today's parents, in addition to paying for the birth and rearing of a child, also bear the cost of education. These are only the direct costs. With the birth of a child, one parent — usually the woman — will tend to spend more time in the home, giving up the income she otherwise would have earned. Depending on her educational background, a woman's loss of earnings over a period of 14 years due to the birth of the first child might be as high as $60,000 to $150,000 (Table 8–2). Young mothers, especially those under 19 years of age, may experience additional costs from early childbearing. These include higher risks in their children of prematurity, mortality, and serious physical and intellectual impairments than among children of mothers 20 to 35 years old. Also, the probability of divorce is considerably higher for couples who married when the wife was 19 years of age or younger.

Sociopsychological Effects of Population Growth

The incidence of mental illness, abnormal behavior, and social pathology is probably higher in crowded populations of mammals. John B. Calhoun of the National Institute of Mental Health raised rats within a laboratory enclosure, and provided

TABLE 8–2   The Total Cost of a Child, 1969*

	Discounted	Undiscounted**
Cost of giving birth	$ 1534	$ 1534
Cost of raising a child	17,576	32,830
Cost of college education	1244	5560
Total direct cost	20,354	39,924
Opportunity cost for the average woman†	39,273	58,437
Total cost of a first child	$59,627	$98,361

*Source: Reed, R. H., and S. McIntosh: Costs of Children.Prepared for the Commission on Population Growth and the American Future, 1972. From Population and the American Future, 1972.

**Discounted and undiscounted costs — spending $1000 today costs more than spending $1000 over a 10-year period because of the nine years of potential interest on the latter. This fact is allowed for in the discounted figures by assuming interest earned annually on money not spent in the first year. True costs are not accurately reflected in the undiscounted estimates, for these are simply accumulations of total outlay without regard to the year in which they must be made.

†Depending on the educational background of the mother, the opportunity costs (earnings foregone by not working) vary.

them with an abundance of food and shelter, in order to study the effects of crowding on their behavior. As the population multiplied and became crowded, he found increasing instances of abnormal behavior. Stress from social interaction and pathological "togetherness" led to disruption of the normal activities of courting, nest-building, nursing and care of the young so that under crowded conditions few young survived. Many females were unable to carry pregnancy to full term or to survive delivery of their litters if they did. Infant mortality ran as high as 96 per cent. Male overactivity led to pathological withdrawal. In short, Calhoun found that many of the normal behavioral activities of rats break down under social pressures generated by increasing population density (crowding).

Rats are not the only animals adversely affected by high densities of population. Studies indicate that at high population densities wild monkeys experience a general breakdown of social order and develop extremely aggressive behavior, hypersexuality, killing of their young, and other maladaptive behaviors. Thus, although all species do not react identically to crowding, high population density appears to have serious inhibiting effects on many animals.

Omar R. Galle et al. at Vanderbilt University studied the relationships of population density and social pathology among humans. They examined statistically the correlations between human population density in 75 community areas of Chicago and a variety of pathological behaviors. As measures of human social pathology they used indices of fertility, mortality, ineffectual care of young, asocial aggressive behavior, and psychiatric disorder. As a measure of ineffectual parental care of the young, they used the number of recipients of public assistance under 18 years old. In measuring asocial aggressive behavior, they used the number of male individuals between ages 12 and 16 brought before county courts on delinquency charges. Using multivariate statistical analysis they subtracted out any effects on these pathologies resulting from differences in social class or ethnic background. They found that usual simple measures of population density (persons per hectare or per square kilometer) may be insufficient in analyzing the effects of crowding. A more significant measure of density and crowding at the personal or individual level may be found in "interpersonal press." They analyzed the relationship between population density and pathology by breaking down density into four component parts: (1) the number of persons per room; (2) the number of persons per housing unit; (3) the number of housing units per structure; and (4) the number of residential structures per acre. With the exception of admissions to mental hospitals, the number of persons per room had the greatest effect in increasing social pathology and "abnormal" behavior.

BIRTH CONTROL

What Is the "Right" Number of People?

In December, 1969, the American Association for the Advancement of Science held a symposium in Boston to consider the question: Is there an optimum level of population? Most people in the United States are aware of crowding, pollution, and other consequences of large urban populations and do not doubt that overpopulation causes serious problems. On the other hand, if the population is too small, it cannot provide all the services necessary for a high quality of life, such as cultural facilities, hospitals, and so forth. "Optimum" population depends on one s interests, and population levels considered too high by one person may be considered too low by another. Some scientists now believe that the United States has already reached or even passed its optimum level of population. They state that in addition to cultural and societal outlets, human beings require wilderness, out-of-doors, and privacy, commodities which are rapidly becoming difficult to find.

Prior to the 1969 meeting in Boston, Fred Singer, deputy assistant secretary of the Department of the Interior, defined optimum population as follows:

. . . the situation in which the population, as a whole, enjoys the highest quality of life. This means, of course, that each person receives an adequate amount of food; is adequately supplied with the necessary raw materials to make the things and devices he needs (including such nonrenewable resources as metals); that there is an adequate supply of energy, as well as water and air of high quality. But, in addition to the so-called "necessities" of life, there are other requirements: adequate medical care to insure good health; recreational facilities, especially outdoors; and cultural outlets. Then there are sociological and psychological requirements, including a requirement for space and privacy.

We must make a choice about future population growth, a choice which has the goal of improving the quality of each life that is actually lived. The Commission on Population Growth and the American Future concluded that the impact of population growth on the economy, resources, environment, government, and society at large indicates the desirability of a slower rate of growth. Continued population growth cannot promote a higher quality of life either in the United States or in the world as a whole. Modern birth control technology has been heralded as the means of stopping this human population growth.

Will Modern Birth Control Technology Solve the Population Growth Crisis?

Birth control is nothing new—it has probably been practiced by man for thousands of years. In fact, James B. Neel, professor of Human Genetics at the University of Michigan, believes that primitive man's development of population control may have been as important as the development of speech and tool making in the transition from higher primates to man. No other primates purposely limited their population size. Natural enemies and the harshness of the environment limited explosive growth of their populations. In man, on the other hand, human social organization and parental care permitted the survival of more children than the culture and economy could support. Thus, early man practiced population control including abortion and infanticide to limit the size of the family. Neel believes the extreme slowness of human population growth until 10,000 years ago resulted not primarily from high infant and childhood mortality caused by infectious and parasitic diseases, but rather from effective limitation of the natural live-birth rate by primitive peoples. This was certainly not done for the purpose of population control per se but may have evolved culturally as a means of survival in the face of limited resources.

Controlled growth of the world population must come about through a reduction of the birth rate by family planning and contraception (Fig. 8–10). However, a 1971 study by Van Tienhoven et al. indicates that widespread ignorance of contraceptive techniques exists even among the most highly educated segments of society. In a survey of 776 men and 278

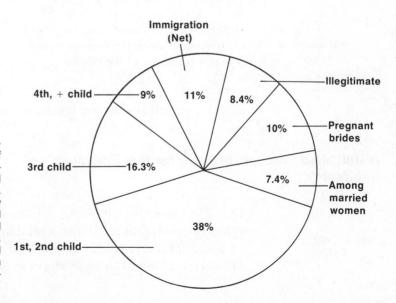

Figure 8–10 Where is US population growth coming from? If additions to the American population were cut in half we would have zero population growth. This would be accomplished now if we could eliminate all unwanted pregnancies and all 3rd, 4th, and additional wanted pregnancies. [From Planned Parenthood Center of Seattle, 1971. Data from 1970 census and Westoff (Princeton) study.]

Immigration (Net)

Illegitimate

4th, + child — 9%   11%   8.4%

10% — Pregnant brides

3rd child — 16.3%

7.4% — Among married women

38%

1st, 2nd child

women (students, graduate students, and faculty at Cornell University), they found much misunderstanding, especially concerning the consequences of sterilization.

Voluntary sterilization, in particular vasectomy for the male, is a simple and relatively safe operation. However, when asked whether or not a vasectomy (male sterilization) would eliminate the ability to ejaculate, a total of 47 per cent of respondents said either that it would (17 per cent), probably would (12 per cent), or that they did not know whether it would (17 per cent). Even the biologists on the faculty showed a high percentage (22 per cent) of ignorance with respect to vasectomy, and 14 per cent of the biology faculty expressed certainty that vasectomy would lead to an inability to ejaculate; 10 per cent were certain that severance of the female oviducts would affect the menstrual cycle, another falsehood.

The majority of the people in the United States probably favor some form of family planning and contraception. Nevertheless, family planning and contraception alone will not solve the population growth problem. The Cornell study found only relatively few (3.7 per cent) were opposed to limiting family size or were uncertain (12.6 per cent) as to whether they favored it or not. However, they disconcertingly found that most respondents professed a preference for families of relatively *large* size. Sixty-five per cent said they wanted three (39 per cent) or more than three (26 per cent) children. Even graduate students in biology included as a group as many as 50 per cent who desired three or more children. It seems clear, as the study points out, that many people consider that "the population bomb is everybody's baby'—except their own.

Because most people desire more children than is good for society as a whole, basic changes in attitude are needed to reduce the growth of population. As Van Tienhoven states:

> To us the deficiencies are clear: our young have yet to learn all that needs to be known about the biology of sex, and they remain yet to be persuaded of the need for reproductive restraint.

Toward this end the reader is referred to Table 8–3.

Is Unlimited Freedom of Reproduction Ethical or Desirable?

The question of limiting population growth brings forth ethical considerations. Daniel Callahan, director of the Institute of Society, Ethics and the Life Sciences, reviewed some of the ethical issues posed by population limitation and proposed

some solutions. Ethical questions arise because limitation of
reproductive freedom challenges existing or desired values. In
the western world, the three dominant ideals are freedom, justice, and security-survival. In proposing methods to limit population growth we are forced to rank or evaluate these three values. Must we give up the freedom to reproduce in order to
insure our very security and survival? Must we give up some of
our security in order not to sacrifice justice? Can we tolerate
reproductive freedom if it jeopardizes justice? The United Nations declaration on human rights states that ". . . couples have
a basic human right to decide freely and responsibly on the
number and spacing of their children and the right to adequate
education and information in this respect." However, relying
on the conscience of individuals to limit the size of their families
is not likely to succeed as a method of population control. Most
families desire more children than could possibly be beneficial
to society in the long run.

Reproductive behavior will have to be limited by social
sanctions against large family size. Individuals will not generally curtail action that is immediately beneficial to themselves in
order to gain a potential (and usually invisible) benefit for society. As Garrett Hardin implies, birth control should be implemented by "mutual coercion, mutually agreed upon by the majority of the people affected." This is not a new concept. Society
has agreed in many cases to "mutually coerce" individuals away
from actions deemed undesirable and to limit the freedom of
individuals for the good of society as a whole; laws punish individuals for stealing money, or prohibit them from parking in
unauthorized areas. Such sanctioned "mutual coercions" are
not always enjoyable—individual freedoms may be limited by
the payment of taxes or by compulsory military service.

Governments may have to intervene in order to limit individual freedom to reproduce because this freedom is not in the
interest of the common good, just as they now intervene to limit
other socially undesirable behavior. In fact, governmental intervention in reproductive behavior is nothing new. There is a
long history of governmental regulation of marriage, sexual behavior, and abortion. Governments probably have as much
right to intervene in reproductive behavior as in other areas
that affect the general welfare. The important issue will be *how*
this governmental right is exercised; namely, governmental
acts (including reproductive limitation) should represent collective democratic decisions. In other words, as Hardin has
suggested, they should represent "mutual coercion mutually
agreed upon."

Governmental actions directed toward limiting population
growth could take many different forms—each with its own
ethical questions. Some authors have suggested such improba-

TABLE 8–3   Birth Control Methods*

	RHYTHM	CONDOM	CONTRACEPTIVE FOAM
What is it?	A plan not to have sexual intercourse during a woman's fertile period — that is, just before, during and after ovulation which is usually from eight to twelve days.	A thin disposable sheath (like the finger of a glove) made of rubber, gut, or plastic. It is worn over the penis during sexual intercourse.	A chemical in a small aerosol container which when shaken becomes spermicidal foam.
How does it work?	Most women release an egg cell once a month — usually 14 days before the next menstrual period. Onset of a period varies from month to month, so it is impossible to be absolutely sure when not to have sexual relations. Women with irregular periods will find this method unsatisfactory.	It catches the man's sperm so they cannot enter the vagina and continue up the woman's reproductive system to fertilize an egg.	It is inserted deep into the vagina with a plastic applicator. The foam is a chemical barrier over the entrance to the uterus. Sperm cells die when in contact with the foam.
How do I use it?	Consult a doctor or family planning clinic for help in determining your fertile period. You will need to record dates of menstrual periods for several months, and perhaps to keep a record of early morning temperature for several months.	It is unrolled over the penis after erection but *before* the penis reaches the vagina and well before ejaculation. After ejaculation, it is removed *away* from the vagina. Then it is thrown away.	You shake the bottle, fill the applicator with foam. Then *lying down* you insert the applicator into the vagina and push the plunger of the applicator. Foam must be inserted no more than a half hour before each sex act.
How reliable is it?	14 to 40 women out of 100 using the calendar method may become pregnant in any given year.	Very few women would become pregnant if their partners used the condom carefully and consistently. Probably less than 3 women out of 100 get pregnant when the partners use a condom, and they also use foam.	About 18 out of 100 women become pregnant in a given year when using foam. If their partners use a condom as additional protection, less than 3 women get pregnant in a given year.
Are there any dangers connected with it?	No.	No.	No.
Does it have any other side effects?	No.	Condoms are an excellent method of preventing veneral disease.	No.
How much does it cost?	If your doctor recommends a thermometer to record early-morning temperatures, it will cost around $5.	25¢–35¢ each at a drugstore. Low cost or no cost for registered patients at Planned Parenthood or other family planning clinic.	It ranges from $2.25 to $3.95 at a drugstore for the whole kit. Low cost or no cost at Planned Parenthood or other family planning clinic.
Where can I get it?	If a special thermometer is needed, it can be bought at a drugstore.	At the drugstore, Planned Parenthood, or other family planning clinic. No prescription is required.	Any drugstore, Planned Parenthood, or other family planning clinic. No prescription required.
How does it affect sex relations?	It hampers spontaneity. Most couples find it difficult to refrain from sexual relations for the length of time needed to be "safe".	If the couple is comfortable enough with each other to make the placing of the condom a part of love-making, it should have no effect. Some men say that when it is in place, it interferes with full sexual pleasure.	Some people do not like having to use a contraceptive before each sex act. Otherwise, no problems.

*By permission of Planned Parenthood Center of Seattle.

Table 8–3 continued on the opposite page.

TABLE 8-3 Birth Control Methods

DIAPHRAGM	INTRAUTERINE DEVICE ("IUD")	"THE PILL"	STERILIZATION
A saucer-shaped device made of rubber stretched over a flexible ring. Inserted in the vagina, it fits snugly over the entrance to the uterus.	Small plastic shapes, some banded with metal, having nylon threads attached to the bottom.	The most common pill combines synthetic hormones, estrogen and progesterone, essentially the same as those created by the ovaries in a woman. Other pills are made of single hormones and are taken in sequence. Pills come in packages sufficient for one month.	Cutting and tying of the tubes in males or females to prevent sperm or eggs from moving into their reproductive system and causing a pregnancy. The male operation is called vasectomy, the female operation is called tubal ligation.
It is used with vaginal cream or jelly, one teaspoonful inside the curve of the diaphragm and one around its stiffened edge. The cream or jelly kills sperm, and the diaphragm keeps them from entering the uterus.	The proper style is inserted into the uterus by a doctor and left there indefinitely. It prevents an egg from being implanted in the wall of the uterus.	The hormones prevent an egg from being released from either of the woman's ovaries. With no egg in a Fallopian tube ready to be fertilized, a woman cannot become pregnant.	In a vasectomy, the tubes between where a man's sperm are produced and where the seminal fluid is made are cut and tied. Afterwards, the sperm made in his testicles cannot move up into his body to get into his semen. He can still enjoy sex and have orgasm with ejaculation but he cannot make a woman pregnant. In a tubal ligation, the Fallopian tubes between a woman's ovaries and her uterus are cut and tied, or cauterized. Afterwards, the eggs which ripen in her ovaries do not pass through the tubes into her uterus. Because no eggs are in the tubes to be fertilized by sperm the woman can't become pregnant.
A diaphragm must be fitted to you by someone medically qualified, who will show you how to insert it (size must be checked every two years and after each pregnancy). Insertion may be as long as six hours before intercourse. If longer, or if a couple wants intercourse again, more cream or jelly must be inserted in the vagina. The diaphragm must be left in at least six hours after intercourse. On removal, it should be washed and dried.	It is inserted by a doctor. After insertion a woman should check after each menstrual period by feeling for the nylon threads which extend down from the device in her uterus into her vagina. If they are there, the device is in place.	Ask your doctor, Planned Parenthood, or other family planning clinic. The usual schedule is one pill per day for 20 or 21 days, starting with the fifth day of a menstrual period. When the pill package is empty the woman stops the medication. In a couple of days she will begin menstruating, and start taking the pills from a new package on the fifth day of the new period. Some packages contain 28 pills, but the last seven contain no medication.	Consult a doctor, Planned Parenthood, or other family planning clinic.
3 to 5 women out of 100 in a given year may become pregnant when using a diaphragm consistently.	2 to 4 out of 100 women using IUDs for a year may become pregnant.	Of 100 women who take the pill regularly only 0.7 to 1.4 may become pregnant. It is more effective than any other birth control method except sterilization.	Virtually 100% sure. No one should be sterilized unless he or she is sure he has had all the children he wants.
No.	After insertion of an IUD, there is a small chance of infection due to germs in the vagina. In case of severe pain or bleeding the doctor may remove the IUD. In a very few cases, perforation of the uterus has been known to occur.	Pills may not be prescribed for some women. Rarely, abnormal blood clotting may occur which can cause serious health difficulties. They are not indicated for women with liver disease, cancer of the breast, or unexplained vaginal bleeding.	There is *some* risk to all surgery, but modern techniques of vasectomy and tubal ligation are very safe indeed.
No.	Many women adapt easily to an IUD. Others — particularly those who have not had children — have cramps and heavy menstrual bleeding, apparently caused by the body's effort to expel the device. Heavier menstrual periods are to be expected the first few months.	In some women it causes weight gain due to water retention; it causes some to feel nauseated. On the positive side, it clears up many women's acne, makes their menstrual periods more regular, makes menstrual flow less heavy, and may minimize cramping.	None.
At a drugstore, diaphragms range from $3.00 to $5.50, which does not include the doctor's fee for fitting. Low cost or no cost at Planned Parenthood or other family planning clinic.	Since the doctor inserts the IUD in his office at the time of the appointment, he charges for the appointment, the device, and a follow-up visit. The cost varies from $20.00 to $50.00 for all three. Low cost or no cost at Planned Parenthood or other family planning clinic.	Costs range from $2.25 to $3.00 for a month's supply from a drugstore. Low cost or no cost at Planned Parenthood or other family planning clinic.	Vasectomy: $50 – $150 Tubal Ligation: Right after childbirth: $125 – $250. Otherwise, as above, plus hospital stay, where applicable. Special care can often be arranged for low income families through Planned Parenthood.
Any drugstore, Planned Parenthood, or other family planning clinic.	From a doctor, Planned Parenthood, or other family planning clinic.	From a doctor, Planned Parenthood, or other family planning clinic.	A vasectomy can be done in a doctor's office. Tubal ligations require a hospital stay of a few hours or some for a day or more.
If properly fitted and inserted, it is not felt by the man or the woman during intercourse. Some women do not like to insert and remove it, which may affect their attitude toward sexual relations.	It does not, except by relieving anxiety over possible pregnancy. When properly inserted, it is felt by neither partner.	It does not, except perhaps indirectly by relieving anxiety over possible pregnancy.	Most patients report satisfaction with the operation. Many feel that removing fear of pregnancy has enhanced sexual pleasure.

ble measures as addition of mass fertility control agents to the water supply, licensing the right to have children, or compulsory sterilization or abortion. Negative incentive programs could take the form of withdrawal of child family allowances and tax deductions or maternity benefits after a given number of children (see box insert). Positive incentives might be to provide people with money or goods in return for "voluntary" regulation of their fertility. These would include financial rewards for sterilization (as is currently done in India), for the use of contraceptives, or for periods of nonpregnancy. Other government policies against population growth could include raising the permitted age for marriage or supplying monetary bonuses for delayed marriages.

OVERPOPULATION AND GOVERNMENT CONTROL

It is 1980. Sara and Eric Nelson just had their third child. She didn't really feel it was right to have more than two, but he had so wanted four that she decided to compromise.

Now they're both paying for it—literally. The government reversed its economic policies back in 1975. The Nelsons now have to pay taxes on the third child, pay for public schooling for the third child, even pay usage fees on facilities usually public—but again for just the third child.

The first two are exempt from all taxes and fees.

And because of that third child, they are no longer eligible for government housing or government loans or government maternity benefits or government scholarships or government pensions, or . . .

With only two children, they would have kept their eligibility.

R. Wright
Christian Science Monitor
July 5, 1972

Any program of planned population control must consider the genetic consequences of limiting the reproduction of segments of the population. James B. Neel proposed the following recommendations with regard to the immediate future of modern human genetics coupled with population growth limitation: *First*, the stabilization of the size of the gene pool by establishment of a quota system of three living children per couple (the failure to marry, infertility, and voluntary limitation to less than three would result in a realized average of approximately two children per family who reach the age of reproduction). Provision would be made for voluntary sterilization after the third child. *Second*, the gene pool should be protected against damage (deterioration) resulting from increasing exposure to environmental mutagens such as radiation, chemicals, and other industrial by-products that could increase the frequency of harmful mutations. "A society that can afford to send man to the moon surely has the resources and the intelligence to monitor itself properly for increased mutation rates." *Third,* the quality of life and the quality of the gene pool should be improved through parental choice based on genetic counseling and prenatal diagnosis (see Chap. 5). And *fourth,* the outward manifestation of individual genetic constitutions should be improved; that is, society should provide greater opportunities for expression, especially for genetically gifted individuals.*

As populations grow and crowding conditions increase, individual freedoms necessarily become less, and individual reproductive freedom may no longer be tolerable in the face of emerging overpopulation. By exercising unlimited individual reproductive freedom, couples can only curtail the future freedom of others. In fact, in the long run, an important result of continued population growth may be a *loss* of individual liberty. The slow erosion of freedom resulting from such unrestricted growth occurs in many subtle ways, as pointed out by the Commission on Population Growth and the American Future:

> Freedom from public regulation — virtually free use of water, access to uncongested, unregulated roadways; freedom to do as we please with what we own; freedom from permits, licenses, fees, red-tape, and bureaucrats; and freedom to fish, swim, and camp where and when we will. Clearly, we do not live in this way now. Maybe we never did. But everything is relative. The population of 2020 may look back with envy on what, from their vantage point, appears to be a relatively unfettered way of life.

If present population growth continues, the resulting crowded world will have no freedom whatsoever. We must sacrifice our right to unlimited reproduction in order to retain other freedoms, and to insure the very survival of future generations. At the same time we should carefully analyze specific proposals for population limitation in order to guarantee that they represent mutual coercion of a majority of the individuals involved and that they include the greatest degree of justice for various segments of the population. In the near future it may become the difficult task of government to examine birth control policies by weighing its mutual tasks of protecting individual freedom on one hand and promoting the collective welfare or "common good" on the other.

---

*See Chapter 5 for a discussion of the possible dangers of such genetic engineering.

REFERENCES

Bajema, J.: The genetic implications of population control. Bioscience *21*(2):71–75, 1971. (Stresses importance of coupling a eugenic program with population control to insure both the quality and quantity of future human population.)

Calhoun, J. B.: Population density and social pathology. Sci. Am. *206*(2):139–148, 1962. (Populations of lab rats when crowded develop abnormal behavior.)

Callahan, D.: Ethics and population limitation. Science *175*:487–494, 1972. (What ethical norms should be brought to bear in controlling population growth?)

Davis, K.: Population. Sci. Am. *209*(3):62–71, 1963.

Deevey, E. S., Jr.: The human population. Sci. Am. *203*(3):194–204, 1960. (History of world population growth.)

Duncan, B.: Regulating human fertility. Science *175*:743, 1972. (A book review of *Rapid Population Growth* by the National Academy of Sciences.)

Ehrlich, P. R.: The Population Bomb. New York, Ballantine Books, Inc., 1968 (paperback). (The magnitude of world population growth and the importance of halting it. "Population control is the only answer.")

*Ehrlich, P. R., and A. H. Ehrlich: Population, Resources, Environment: Issues in Human Ecology. San Francisco, W. H. Freeman and Co., 1970. (An excellent introductory text on population, world resources, pollution, birth control, etc.)

Frejka, T.: The prospects for a stationary world population. Sci. Am. *228*(3):15–23, 1973. (General discussion of world demography and different projections for future population growth.)

Galle, O. R., W. R. Gove, and J. M. McPherson: Population density and social pathology: what are the relations for man? Science *176*:23–30, 1972.

*Hardin, G.: The tragedy of the commons. The population problem has no technical solution; it requires a fundamental extension in morality. Science *162*:1243–1248, 1968.

Hardin, G., ed.: Population, Evolution, and Birth Control: A Collage of Controversial Ideas. San Francisco, W. H. Freeman and Co., 1969. (Collected essays.)

Kangas, L. W.: Integrated incentives for fertility control. Science *169*:1278–1283, 1970. (Proposes wider use of positive material incentives for family planning.)

Langer, W. L.: The black death. Sci. Am. *210*(2):114–121, 1964. (History of the plague epidemic and its effect on population growth.)

Malthus, Thomas R.: 1798. An Essay on the Principle of Population. Homewood, Illinois, Richard D. Irwin, Inc., 1963 (paperback). (The classic and early statement of the problem of human population growth.)

Mudd, S., ed.: The Population Crisis and the Use of World Resources. Bloomington, Indiana, Indiana University Press, 1966.

*Population and the American Future. New York, The New American Library (Signet), 1972 (paperback). (The report of the US Commission on Population Growth and the American Future.)

Population Reference Bureau, Inc. Washington, DC, 20036. Many publications available, for example: No. 37, Perspectives from the Black Community; No. 30, Social Consequences of Population Growth; No. 29, Population Resources and the Great Complexity.

Singer, S. F.: Is there an optimum level of population? Science *166*:270–271, 1969.

Singer, S. F., ed.: Is there an optimum level of population? New York, McGraw-Hill, 1971. (AAAS Symposium, Boston, December 1969. Symposium of scientists with widely differing views on the question.)

Van Tienhoven, A., T. Eisner, and F. Rosenblatt: Education and the population explosion. Bioscience *21*(1):16–19, 1971. (Study indicates that even highly educated segments of American society need further instruction on population growth and birth control.)

Zero Population Growth: A Teacher's Guide to Materials on Population. Denver, Colorado, ZPG, PO Box 18291, 1971. (An excellent source of further materials on population and resources. Includes bibliography of books, audio-visual materials, and·population games.)

*Recommended further reading.

PROBLEMS

1. What is being done to curb population growth in areas where the rate is now high?

2. Do researchers believe that making contraceptives available to everyone upon request would lessen the population bomb? Would it accomplish zero population growth?

3. Which countries are concerned about or doing something about overpopulation?

4. Why do many couples want more than two children?

5. Will educating people about the problems of overpopulation and then letting them decide for themselves how to solve these problems lead to a decrease in the birth rate?

6. Will a family size of two children per family (allowing for deaths) eventually stabilize or decrease the population? Why?

7. Should population control be directed by individual countries or internationally? If the latter, how can we go about setting up such an international body to govern population growth?

8. Do you believe there should be a government committee to control the number of children a family can have? Do you feel that their decision should be final, or should there be another committee to which families could appeal the decision?

9. What present day laws actually give some people an incentive to increase their families?

10. Should an affluent country make the giving of food or aid to its neighbors dependent upon establishment of a population control program in the poorer country?

9

WORLD
RESOURCES

*Every seven-and-a-half seconds a new American is born. He is a
disarming little thing, but he begins to scream loudly in a voice that
can be heard for 70 years. He is screaming for 26 million tons of water,
21,000 gallons of gasoline, 10,150 pounds of meat, 28,000 pounds of
milk and cream, 9000 pounds of wheat, and great storehouses of all
other foods, drinks, and tobaccos. Those are his lifetime demands on
his country and its economy.*

Robert and Leona Rienol

The necessities of life must no longer be taken for granted.
The first symptoms of the environmental crisis predicted by
some ecologists may be emerging at this very moment — these
symptoms are reflected in the growing frequency and magnitude of resource shortages. The quality of life for many of us is
now being challenged by real shortages in *water, food, minerals,*
and *energy*. Can man, through science and technology, meet this
challenge and continue to supply these necessities, or will he
face ever-increasing shortages with a consequent lowering of
the quality of life? What do we know about present levels of
resource consumption? Can we meet projected future demands
for these four essential resources?

WATER

Water is necessary to sustain all forms of life. In the United
States, which undoubtedly has the world's largest per capita
water use, consumption can be divided as follows: about 8 per
cent for domestic and personal needs, 46 per cent for agricultural irrigation, and the remaining 46 per cent for industrial

257

TABLE 9-1 Water Requirements in Industry*

The Manufacture of One Ton of:	Requires the Following Tons of Water
Synthetic rubber	2500–3000
Coal	2250–2750
Newsprint	900–1000
Paper pulp	175–250
Steel	160–260
Gasoline (from coal)	500–900
Oil (refining)	30–60

*From Borgstrom, 1967, p. 423.

uses. Large quantities of water are needed for each of these purposes. For personal use alone, each United States urban inhabitant averages 473 to 566 liters of water per day for drinking, personal hygiene, sanitation, lawn watering, and air conditioning. If we add to this the large amounts of water required by industry to produce the manufactured goods used by each person, the *real* use of water in the United States totals about 5000 to 8000 liters per person per day. Russian estimates indicate that each person in a modern technical society requires at least 2700 liters of water per day.

Water is essential for food production. For example, 580 to 780 liters of water are required to produce one kilogram (kg) of potatoes, and about 132 liters of water are required to produce one slice of bread (including the production of wheat grain in the field). Meat production requires even larger amounts of water. If we include not only the water drunk by cattle or ingested in their food, but also that required in the field to produce feed grain, beef cattle consume about 32,000 liters of water for each kg of meat produced.

Huge amounts of water are required for industry, much of it used for cooling in the production of both goods and energy (Table 9-1). To produce one automobile requires directly and indirectly about 378,000 liters of water. The increasing shift to the production of synthetic goods, which require even larger amounts of energy than natural products, is accompanied by a need for more water in industry. The production of synthetic fibers requires between 2500 and 3600 liters of water per kg of fiber; one kg of cotton fabric, on the other hand, requires only about 7 liters of water to produce.

Even though water circulates from the land to the sea to the atmosphere and then back to the land, the world's supply of fresh water remains finite. The problem consists not only of supplying the quantity of water to areas where it is needed but also of supplying *quality* water. Water quality in many areas is being threatened by water pollution (see Chap. 7).

The Water Shortage

As population continues to increase, many scientists predict that the world will face an extremely grave water crisis in the near future. Water requirements already exceed available flow in the southwestern United States, and projections indicate that population and economic growth will cause the area of water shortage to expand northward and eastward within the next few decades. Ground water reserves are being depleted in many areas at a shocking rate, and the President's Commission on Population Growth and the American Future estimates that water deficits will continue to grow in the United States despite large expenditures on water treatment, dams, and reservoirs during the next 50 years (Fig. 9–1). Water usage may have to be reduced through rationing and the adoption of water-conserving technologies.

United States water use has increased rapidly in recent years; in many areas ground water must now be pumped from

Figure 9–1   A, By permission of John Hart and Field Enterprises Inc.

Regional Water Shortages
Billions of liters/day

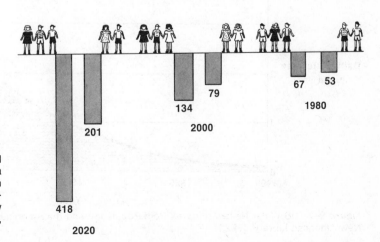

Figure 9–1 B, Projected regional water shortages for the US under a two- or three-child family growth rate to the year 2020. (From Population and the American Future. New York, The New American Library, 1972.)

deep wells for use in agricultural irrigation. In other regions, ground water reserves are becoming depleted. Owing to both the removal from deep wells and to accelerated run-off, ground water levels in Arizona and much of southern California have fallen more than 30 meters below earlier levels.

Strict rationing of water has already been implemented in many places, including some United States urban areas, and in other parts of the world. In several large Latin American and Middle Eastern cities the water is shut off for several hours daily and in some sections for the entire day. Crowded cities like Hong Kong and Tokyo, where lack of water could have catastrophic consequences, recently faced the necessity for water rationing. As water supplies become scarcer, water will have to be recycled and used repeatedly, and costly safeguards will be required to avoid pollution of the limited water available. Water in many areas of the Ohio River Basin is already being utilized many times over. The total quantity of water available from all sources in the United States is estimated at 2.1 $\times 10^{12}$ liters. This level of use will probably be reached sometime between 1975 and 1980 (Fig. 9–2).

Desalination. There is some optimism that (at least at increased cost) fresh water can be supplied through desalination of the ocean and other bodies of salt water. One-third of the world land area is essentially uninhabited owing to inaccessibility to fresh water, and large areas of this dry land, especially near the ocean, could be used for human settlement if fresh water were available. Thus, in recent years interest in the large-scale desalting of seawater has grown rapidly, and pilot plants have been established to test the feasibility of producing fresh

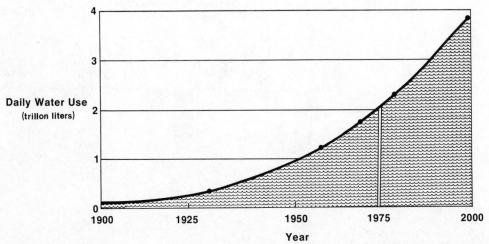

Figure 9–2 US Water Needs. (Figures from Population and the American Future. New York, The New American Library, 1972.)

water from seawater. The most promising of these plants are dual purpose plants producing both power and water. The oil producing country of Kuwait on the Persian Gulf and the community of Shevchenko on the barren eastern shore of the Caspian Sea now depend on desalination for their water supply. The importance of desalination is stressed by Jacob Bronowski, who says that "over the next 20 to 30 years . . . desalting of seawater is going to be a most important advance for overall world development."

Can fresh water be produced from seawater at a competitive price to be used in the large quantities necessary for irrigation farming? Gale Young, assistant director of the United States Oak Ridge National Laboratory, believes that desalination agriculture is a real possibility for the near future. Research indicates that cultivation of rice — the preferred staple of half the human race — does not require much more water than other grains. With improved desalination technologies, rice could be grown with desalinated water at prices competitive in the world market.

Desalination of seawater is often proclaimed as the solution to the world water crisis, but not all agree on its feasibility. Irrigating the southern part of Texas alone would require 15,000 large desalination plants and expensive pumping and piping of water over hundreds of miles. Furthermore, a profitable use would have to be found for the huge amounts of salt extracted by desalination — the manufacture by desalination techniques of one ton of milk, for example, would produce more than 22 metric tons of salt! George Borgstrom pessimistically concludes, "Proposals for irrigation on a continental scale based on desalination of water are loose speculation. They are most unrealistic in not taking into account the magnitude of the water needs for food- and feed-producing crops."

We must not only maintain our supply of water but also the *cleanness* of the water. A combination of pollution and certain viruses resistant to chlorine treatment are beginning to threaten the quality of ground water supplies, and in some areas doctors are prescribing pure bottled water for infants, in an effort to protect them from harmful agents in ground water.

FOOD

A second essential requirement for an acceptable standard of life is adequate food. If human beings do not receive enough food energy in calories they become *undernourished;* if they do not receive the minimum levels of vitamins, minerals, or protein, they become *malnourished.* Pregnant women and growing children require more protein than the average person. Dwarf-

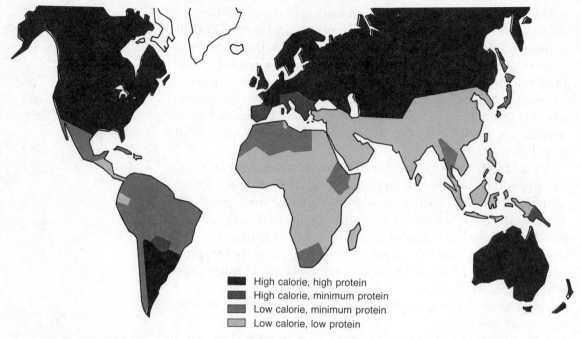

Figure 9-3 The geography of hunger. (From Ehrlich, P. R., and A. H. Ehrlich: Population, Resources, Environment: Issues in Human Ecology. Second edition. San Francisco, W. H. Freeman and Co., copyright © 1972.)

ing, delayed physical maturity, mental impairment, and lowered resistance to disease can all result from protein deficiency. At present, about 20 per cent of the world's population (mostly in the underdeveloped countries) is undernourished, while 60 per cent are malnourished (mainly in proteins). United Nations data indicate that in many countries protein consumption does not reach the required average level of 58 grams per day. For example, the average Indian receives only 50 grams and the average Filipino 46 grams of protein per day. In fact, two-thirds of the world's people do not receive a sufficient diet (Fig. 9-3).

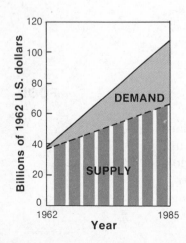

Figure 9-4 Food gap is caused by demand, measured in 1962 US dollars, which is rising faster than supply. Unless developing countries increase food production more rapidly than they have, demand for food will exceed supply by about $43 billion in 1985. (Adapted from Boerma, A. H.: A world agricultural plan. Sci. Am. 226(1):70–77, 1970.)

An immediate doubling of the world food output is needed to supply an adequate diet for the people of the world's under-developed areas. By the year 2000, food production will have to increase by two to three times the present rate in order to meet the needs of the predicted doubled population. The question then arises, can we feed these growing billions? Projections indicate that population will continue to grow faster than food production for the next few decades (Fig. 9–4). As R. R. R. Brooks says,

> The key to the . . . future of two-thirds of the human species is rising productivity in agriculture. All political dogmas, party slogans, planning strategies, and models of economic growth shrivel to irrelevance in the face of this fact.

Can We Increase Food Production to Meet Future Demands?

Stirling Hendricks outlines five possible methods of increasing the world's food supply: (1) cultivation of new land, (2) increasing the production of land already in use, (3) prevention of crop losses, (4) radically new technological innovations, and (5) increased economic investment in agriculture.

One fact stands out clearly—the best agricultural lands are already being used. Only about one-fourth of the world's 32 billion acres is potentially arable, and another one-fourth could be used for grazing. At least one-half of this arable land (about one acre per person on a world average) is already being used. Furthermore, in Europe and Asia, the most highly populated areas in the world, approximately 80 to 100 per cent of potentially arable lands are already in use (Table 9–2). Large areas of potentially arable land remain in the tropics, but in these regions heavy rains often leach away necessary plant nutrients from the soil. Agriculturalists are learning that the crops and cultivation techniques of temperate regions are often unsuitable to the tropics.

Agricultural land becomes scarcer and food prices are driven up as the growing population removes more land from use for food production and uses it for housing, industry, and blacktop.

In theory, improvements in the use of land could increase food production, but such increases involve large energy inputs in the form of fertilizer, pesticides, and water, and many practical problems of economics and transportation arise. The industrialized countries use huge tonnages of fertilizers not available to underdeveloped countries. For example, in the United

TABLE 9–2 **Population and Cultivated Land on Each Continent, Compared with Potentially Arable Land***

Continent	Population in 1965 (Millions of Persons)	Area in Billions of Acres			Acres of Culti- vated** Land per Person	Ratio of Culti- vated** to Potentially Arable Land (%)
		Total	Poten- tially Arable	Culti- vated**		
Africa	310	7.46	1.81	0.39	1.3	22
Asia	1855	6.76	1.55	1.28	.7	83
Australia and New Zealand	14	2.03	.38	.04	2.9	2
Europe	445	1.18	.43	.38	.9	88
North America	255	5.21	1.15	.59	2.3	51
South America	197	4.33	1.68	.19	1.0	11
USSR	234	5.52	.88	.56	2.4	64
Total	3310	32.49	7.88	3.43	1.0	44

*From The President's Science Advisory Committee, *The World Food Problem,* 1967.
**Cultivated area* is defined by the Food and Agriculture Organization of the United Nations as "arable land and land under permanent crops." It includes land under crops, temporary fallow, temporary meadows, for mowing or pasture, market and kitchen gardens, fruit trees, vines, shrubs, and rubber plantations. Within this definition there are said to be wide variations among reporting countries. The land actually harvested during any particular year is about one-half to two-thirds of the total cultivated land.

States between 1962 and 1964 more than 3.6 million metric tons of nitrogen were consumed as fertilizer. Such intensive agricultural processes can contribute to the pollution of streams, lakes, and reservoirs through soil erosion and run-off, which carries sediment, nitrogen, and phosphorus fertilizers, pesticides, and organic matter back to the environment.

Twenty years ago, scientists using systematic programs of plant breeding began to develop high-yield "miracle grains." The resulting so-called green revolution was hailed as the technological solution to the food shortage. However, costs of miracle grains are relatively high, and the high per-acre yields depend upon large inputs of irrigation water, chemical fertilizers, herbicides, pesticides, new storage facilities, and development of new transportation services. Another problem of the miracle grain program was that farmers in the underdeveloped countries had to invest heavily in machinery and equipment for this type of agriculture. Consequently, some scientists now question the long-term value of the green revolution as a solution to the world food shortage.

In the developed countries and in areas of the green revolution, production is heavily dependent on energy (primarily fossil fuel) for machinery, transportation, irrigation, fertilizers, pesticides, and so forth. Such *energy intensive agriculture* uses large amounts of energy in food production. Furthermore, modern food processing may require more energy than the food production itself. United States farming uses more petroleum than any other single American industry.

David Pimentel et al. state that although energy intensive agriculture may lead to a large increase in total food production, energy efficiency in terms of production-to-consumption ratio may actually decrease. For example, in the United States in 1945, 3.70 kilocalories (kcal) of corn were produced for each kcal of energy input. By 1970 this ratio had declined by 24 per cent to 2.82 kcal of corn per kcal input. In fact, food production, contrary to popular belief, is considerably more expensive per kcal of plant product in developed countries such as the United States ($38/1000 kcal) than in underdeveloped countries such as India ($10/1000 kcal). A world energy shortage (see *Energy,* later in this chapter) would have a devastating impact on energy intensive agriculture and overall world food supply.

Various other forms of technological innovation have been suggested as means to increase the food supply. For example, synthetic protein can be produced by bacterial conversion from petroleum products. This method, however, consumes a nonrenewable petroleum resource. For some arid areas near the ocean, food production could be increased through desalination of salt water and irrigation, as discussed earlier in this chapter.

Another method of increasing the supply of food is to decrease the amount lost through pests and crop diseases. Severe losses of this nature have been documented. For example, rats in two Philippine provinces during 1952 to 1954 devoured 90 per cent of the rice, 20 to 80 per cent of the maize, and more than 50 per cent of the sugar cane. New varieties of pesticides based on insect hormones are being developed that are more specific and safer than old forms such as DDT. One of the most profitable ways of increasing available food might be to improve methods of shipping and storage of crops after harvest. Some scientists recommend the domestication of several presently wild animal species as an added source of meat, a goal which could also be achieved by a large expansion of pig and poultry production.

Frances Lappé presents convincing arguments that the most efficient way to increase food availability, at least in the United States, is to switch to a vegetable protein diet. Meat is not in itself a dietary requirement, and the protein that meat supplies can be furnished adequately by a balanced vegetable diet. Lappé points out that in the United States about one-half of all agricultural land is planted in feed crops; and 78 per cent of the grain is fed to animals. This represents 18 million metric tons of protein that could be used directly by man. By feeding the grain to cattle and then harvesting the cattle we reap only 5 per cent of our original protein input owing to loss of energy through the food chain (see Chap. 7).

In the United States alone this loss through conversion from vegetable to animal protein is equivalent to 16.4 million metric tons of protein; 90 per cent of the yearly world protein deficit; or 12 grams per day for every person in the world. It is more efficient to use land that can support crops to feed people rather than livestock. One acre of cereal can produce five times more protein than an acre devoted to beef production; one acre of legumes (peas, beans, lentils) ten times this amount; and one acre of leafy vegetables 15 times more protein than one acre used for cattle. Of course some marginal land not suitable for growing crops can best be used for grazing. Nevertheless, Lappé estimates that reducing United States livestock production by one-half and replacing it with vegetable production could release about 91 million metric tons of grain for human consumption—an amount that would meet the caloric deficit of the "nonsocialist" countries almost four times over. It is also important to consider that supplying underdeveloped countries with such "released" grain protein from the United States would involve immense problems of transportation, storage, and political cooperation.

Harvesting the Sea

If the land will not feed the growing population, perhaps harvesting the sea will solve the food problem. Surely, as many have said, the bountiful sea can provide unlimited supplies of food. Unfortunately, such optimism is unwarranted: the point

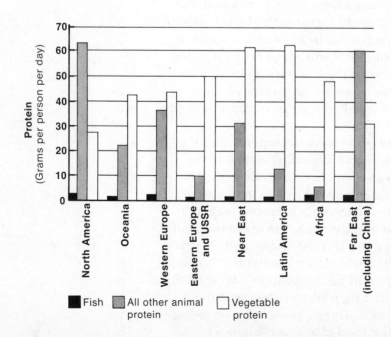

Figure 9–5 The role played by fish in the world's total consumption of protein is small. The grams of fish eaten per person per day in various parts of the world (*left column in each group*) is compared with the consumption of other animal protein (*middle column*). Most of the world's population depends heavily on vegetable protein (*right column in each group*). (Adapted from Holt, S. J.: The food resources of the ocean. Sci. Am. *221*(3):178–194, 1969.)

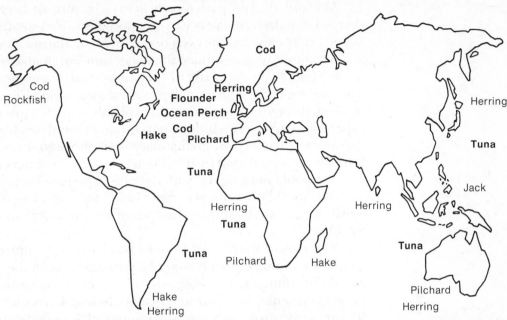

Figure 9-6 Exploitation of fisheries during the past 20 years is evident from this map, which locates 24 major fish stocks that were thought to be underfished in 1949. Today 12 of the stocks (**bold**) are probably fully exploited or in danger of being overfished. (Adapted from Holt, S. J.: The food resources of the ocean. Sci. Am. *221*(3):178–194, 1969.)

of maximal sustained harvest of wild fish stocks from the sea will probably be reached in the near future. Fish constitutes only a small portion of the world's protein supply (Fig. 9–5). Furthermore, increasing amounts of harvested fish are now being used to produce fish meal for animal foods and not used directly for human food. At present, approximately 60 million metric tons of food (chiefly fish) are harvested from the sea each year. Peru, the world's leading fishing nation, has developed a large fishery based primarily on one species of anchovy, of which it harvests about 10 million tons per year (most of which is reduced to fish meal). However, recent evidence indicates this major fishery is now on the decline.

Although the world fish catch has more than tripled in the past three decades, humans consume directly only about half the catch—the rest becomes feed for livestock. The great increases in fish harvests of the past few years are largely the result of technological advances such as efficient factory ships that process harvested fish while still at sea. For example, a single Rumanian factory ship equipped with modern devices caught in one day in the Pacific Ocean as many tons of fish as the whole New Zealand fleet of some 1500 vessels. In fact, many of the world's fisheries have already been fully exploited or even overfished in recent years (Fig. 9–6). To avoid overfishing, depletion, and eventual extinction of species, fish must be harvested on a sustained yield basis; that is, no more should be taken than can be replaced by natural production.

On land, the best agricultural areas have already been developed. Similarly, in the ocean the most desirable food fish are already fully exploited or even overexploited. Marine scientists and oceanographers estimate the maximum sustainable harvest of the ocean to be about 100 to 150 million metric tons per year. The present harvest of 60 million metric tons could be increased, therefore, but this would probably require a greater expenditure of energy and the harvesting of less desirable food species (which is already taking place in some areas). Even if the fish harvest is increased to 70 million tons within the next ten years, it would not keep up with the rate of population growth and thus would not represent a per capita increase. The deficit is even larger when estimated as tons of protein rather than total fish.

S J. Holt of the United Nations Food and Agriculture Organization estimates that it might be possible to reach the world's maximum sustainable fish catch level by 1985 or at the latest by the end of the century. Fishery biologist William Ricker, writing for the National Academy of Sciences, estimates that by the end of this century, even if we reach a production of 150 to 160 million tons, this would supply only about 3 per cent of the energy requirement of the 6 to 7 billion people expected to inhabit the earth at that time. In summary, he concludes, "The world's ocean and inland waters cannot even begin to supply a complete ration for the world's people." Continued successful harvesting of ocean resources will require international agreements, first to limit overfishing and depletion of natural stocks, and secondly to prohibit oceanic dumping of wastes and pollutants. Such oceanic pollution is widespread. For example, scientists recently reported finding large numbers of small plastic particles on the surface of large areas of the Atlantic Ocean. Such forewarnings cause some scientists to speculate that if man does not soon reach international agreements controlling marine pollution, we may live to see the death of the world's oceans.

New marine technologies offer some hope for increasing world food supply. If the organisms and plankton upon which fish feed could be harvested for human consumption, the 100 to 200 million ton maximum potential ocean harvest could theoretically be increased ten times. However, these small organisms are thinly dispersed, difficult to collect, and not very palatable as food. One such organism is the euphausid or *krill*, a small shrimplike crustacean particularly abundant in Antarctic waters, which forms the main food of baleen whales. Russian investigators are experimenting with the harvesting and the development of suitable meals and pastes from the krill.

Mariculture. Probably the most promising technology for increasing ocean harvest is mariculture (controlled marine

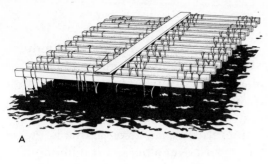

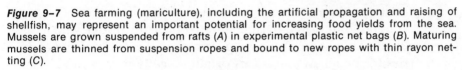

Figure 9–7 Sea farming (mariculture), including the artificial propagation and raising of shellfish, may represent an important potential for increasing food yields from the sea. Mussels are grown suspended from rafts (*A*) in experimental plastic net bags (*B*). Maturing mussels are thinned from suspension ropes and bound to new ropes with thin rayon netting (*C*).

farming of the sea). Mariculture is carried out primarily in shallow coastal waters, where various species of fish, shellfish, and seaweed are planted, grown, and harvested (Fig. 9–7*A* and *B*). However, much basic research remains to be done before mariculture can be expected to produce a large share of the world's marine resources. Furthermore, since mariculture can be most efficiently carried out in only a limited number of shallow coastal waters, and because these are often close to industrial areas, there is an ever-present threat that pollution will destroy marine productivity.

Theoretically harvest from the sea could be increased, particularly through scientifically managed mariculture. However, there is little reason to believe that ocean harvest could grow indefinitely at a rate sufficient to provide for present world population growth.

The greatest promise for augmenting the world food supply probably lies in extending the efficiency of yield from land areas already under cultivation. However, increasing yield means continuing use of new or improved pesticides, intensive fertilization, and energy inputs, all of which contribute to the danger that such intensive agriculture could lead to further environmental deterioration. The world food crisis is uniquely important. In the near future, worldwide supply is almost certain to continue to grow more slowly than demand. Productive, developed countries such as the United States will be forced to sell food in response to international demand and in order to balance their expanded import of minerals and fuels (see later in this chapter). Thus, food prices may continue to rise. In the long run there can be no solution to the world food shortage without simultaneous efforts to control the growth of both world population and pollution.

MINERALS

The third group of resources consumed by modern industrial nations is minerals. The nonfuel minerals currently in greatest demand are probably iron, copper, aluminum, and fertilizer minerals. Other rarer metals such as mercury, however, are also essential to many industrial processes. It is important to realize that these minerals are *nonrenewable;* that is, they were deposited in the earth's crust over a period of millions of years, each deposit is finite, the time required to remove them is very small, and no second crop will materialize within the lifetime of modern civilization. Mineral deposits have been referred to as "quick assets," and countries that possess them ultimately exhaust them. Their essential part in industrial affluence is depicted in the Hewett-Lovering Model, which relates the stages of mineral production of industrialized countries to their national status or affluence (Fig. 9–8).

As with other resources, consumption of minerals has grown significantly in recent years. The quantity of metal consumed by the world since the beginning of World War II is more than that produced during all previous history. The United States is already a heavy importer of metals, and few of the other developed countries have adequate supplies of minerals to extend into the next century, without recycling (Fig. 9–9). The demand for minerals will probably increase exponen-

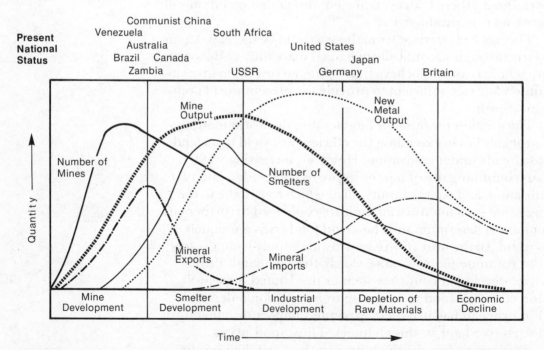

Figure 9–8 Hewett-Lovering model stages of mineral production of an industrialized country. [From Cheney, E. S.: Mining Engineering *19*:12, 47–49, 51, 54, 1967; Adapted from Hewett (1929) and Lovering (1943).]

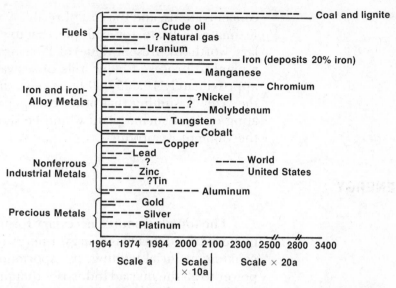

Figure 9–9 Lifetimes of estimated recoverable reserves of mineral resources. Reserves are those that are of high enough grade to be mined with today's techniques. Increasing population and consumption rates, unknown deposits, and future use of presently submarginal ores are not considered. (From Ehrlich, P. R., and A. H. Ehrlich: Population, Resources, Environment: Issues in Human Ecology. Second edition. San Francisco, W. H. Freeman and Co., copyright © 1972.)

tially for at least the next 50 to 75 years before leveling off—and it must level off, because the earth's crust and the minerals it supplies are finite. The highest grade and most accessible ore deposits are the first to be mined. As these rich deposits are depleted and we turn to the mining of poorer deposits, the tonnage of ore necessary and the costs and energy incurred increase geometrically as the quality decreases.

At present, all industrial nations, except possibly the Soviet Union, are net importers of most of the minerals and ores used by them. In the near future, maintenance of supplies in the United States will probably result in a steadily rising cost and increased dependence on importation from other countries, many of which are politically unstable. Some new discoveries of mineral deposits will be made, but these will be mostly in remote and undeveloped areas. Development of recycling processes could tap a currently wasted reservoir of potential mineral resources.

Can we turn to the sea for our mineral requirements? Seawater could theoretically supply most of the manganese, bromine, and common salt that we need. In addition, significant quantities of magnesium, sodium, potassium, iodine, strontium, and boron could probably be extracted from seawater. However, many varieties and large quantities of important industrial minerals present in the ocean cannot be feasibly extracted either from seawater or from the sea floor because most of these minerals are dissolved in seawater in very low concentrations, and the cost of recovery would be prohibitive. Estimates indicate, for example, that to extract 1 million dollars' worth (by present value) of 17 critical metals would require a plant to process 1 cubic mile of seawater per year or approximately 10,625,000 liters per minute every minute of the year. Assuming that other cheaper sources exist, operating costs of such a plant would be significantly greater than the value of its products.

ENERGY

The fourth resource necessary for maintaining an adequate living standard is energy. Energy is necessary for heat, cooking, artificial lighting, transportation, and for supplying power to all the myriad industries that produce the necessities and luxuries that maintain the present quality of life. The ultimate source of the energy we consume is the sun; solar radiation is trapped or stored in various forms and extracted for man's use. The past growth and present affluence of our industrial society has depended largely upon the exploitation of one source of energy, the fossil fuels: coal, natural gas, and oil.

Fossil Fuels

From the year 1700 to the present, the total power output of all world machinery increased about 10,000 times, and we are presently undergoing an exponential rise in energy consumption. In the United States, primary energy sources shifted during that period from wood to coal and finally to gas and oil (Fig. 9–10). Energy consumption during the past 100 years has grown about three times as fast as population. Before World War II, the United States was a net energy exporter. Today the Middle East and Africa possess more than one-half of the world's fossil fuel reserves, and most developed countries, including the United States, must import much of their energy fuels from underdeveloped countries. The United States consumes more than 25 per cent of all the world's fossil fuel production. Of the electricity produced largely from these fuels, the greatest share (40 per cent) goes to industry; the remainder is divided mostly between commercial (22 per cent) and residential (34 per cent) uses. United States energy consumption in 1970 was about 10 kilowatts per capita per year in terms of thermal energy or more than seven times the world average. It has been estimated that world consumption of energy for industrial purposes is doubling approximately every 10 years. The rapid exponential growth in energy consumption becomes evident when we

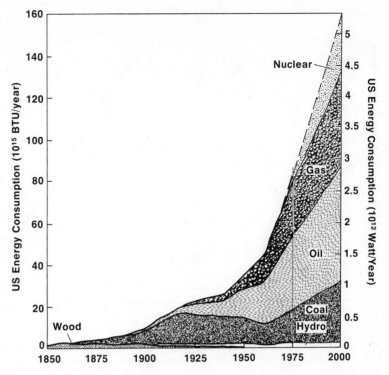

Figure 9–10 US energy consumption has been multiplied some 30 times since 1850, when wood supplied more than 90 per cent of all energy units. By 1900 coal had become the dominant fuel, accounting for more than 70 per cent of the total. Fifty years later coal's share had dropped to 36.5 per cent and the contribution from oil and natural gas had climbed to 55.5 per cent. By 1970 coal accounted for 20.1 per cent of all energy consumed, oil and gas 75.8 per cent, hydropower 3.8 per cent and nuclear energy 0.3 per cent. (Adapted from Starr, C.: Energy and power. Sci. Am. 225(3):37–49, 1971.)

consider that, for example, the quantity of coal produced and consumed in the 35 years since 1940 is approximately equal to the total coal consumption up to that time, and that the amount of petroleum and related products manufactured from 1857 to 1959 is equal to the amount consumed from 1959 to 1969. About 96 per cent of United States energy comes from fossil fuels; petroleum (43 per cent, primarily used for transportation), natural gas (33 per cent), and coal (20 per cent). The remaining 4 per cent is derived from hydroelectric energy (3 per cent) and nuclear energy (about 1 per cent). The fossil fuels coal, petroleum, and natural gas were deposited over a period of millions of years of geological history by a process whereby quantities of vegetable and animal matter were buried under proper environmental conditions of incomplete oxidation and decay. Their supply is finite.

Given the rapidity with which man is consuming them, the petroleum fuels in particular, fossil fuel supplies will soon be depleted. The total length of time that they are available to support industrial affluence is extremely brief. It took more than 600 million years to form most of the petroleum reserves; at present rates of consumption, man will have largely exhausted them within the next 50 years. M. King Hubbert estimates that total petroleum reserves in the coterminous United States and continental shelves are about 165 billion barrels. Discoveries up to 1965 thus represent about 82 per cent of the prospective ultimate total. The peak production for oil, in fact, is occurring at

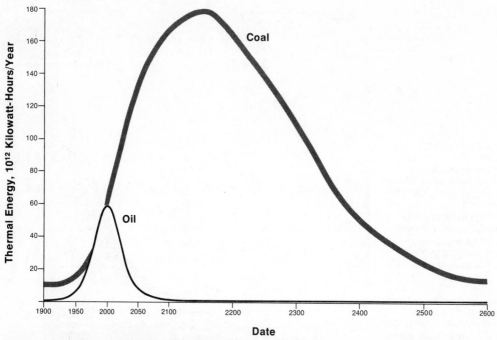

Figure 9–11 Cycle of world coal production (*top*) is plotted on the basis of estimated supplies and rates of production. Cycle of world oil production (*bottom*) is plotted on the basis of two estimates of the amount of oil that will ultimately be produced. (Adapted from Hubbert, M. K.: The energy resources of the earth. Sci. Am. *224*(3):69, 1972.)

the present time, and it is estimated that 80 per cent of the ultimate cumulative total will be produced during the 65-year period between 1934 and 1999 — less than the span of a human lifetime. Oil reserves discovered in the North Slope of Alaska will probably add only about a 10-year supply to the United States at present rates of consumption. For the world as a whole, peak petroleum production is predicted to occur about the year 1990 or 2000, with the largest quantity being produced during the period 1961 to 2032 (Fig. 9–11). As reserves become scarcer and eventually depleted, the cost of finding new reserves and extracting petroleum from them will increase greatly. A substantial but still limited amount of oil could be extracted from tar sands and oil shales but at higher costs. The peak in natural gas production will probably be reached between 1975 and 1980.

Reserves of coal as an energy source extend somewhat further into the future than petroleum reserves. Figures indicate that peak coal production rates will occur, depending upon the rate of extraction, sometime between the years 2100 and 2200 (Fig. 9–11).

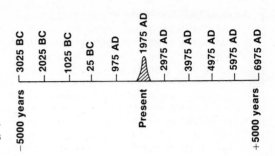

Figure 9–12 Transitory age of practical fossil fuel utilization (1900 AD to 2400 AD) on a time scale from 5000 years ago to 5000 years in the future.

Natural gas currently supplies about one-third of total United States energy requirements, including about one-half of the demand of residential, commercial, and industrial consumers. United States demand for natural gas already exceeds supply, and projections indicate that the gap will increase in the future. Schemes for extracting natural gas from coal and shipping it by pipeline have been suggested as a future source of large amounts of energy (see, for example, Mills, 1971). However, using natural gas produced from coal, although it would probably produce less air pollution than direct coal burning, would not be without its environmental costs. Large areas of strip mining would be necessary to supply the required coal, huge quantities of water would be needed for cooling purposes, and subsequent thermal pollution (Chap. 7) would be a problem.

Thus, exhaustion of 80 per cent of the total original supply of petroleum fuels (crude oil, natural gas, tar, and shale oil) will probably take only one century and exhaustion of 80 per cent of the world's total coal resources about three to four centuries (providing coal is not used as the main energy source). However, there is great uncertainty regarding the accuracy of such estimates. Discovery of large untapped reserves, changes in future use rates, and new technological developments could alter these projections by orders of magnitude. The transitory state and short time span of fossil fuel utilization can best be represented within a historical perspective between 5000 years in the past and 5000 years in the future (Fig. 9–12). As M. King Hubbert points out, "On such a time scale, it is seen that the epic of the fossil fuels can only be a transitory and ephemeral event—an event, nonetheless, which has exercised the most drastic influence experienced by the human species during its entire biological history."

THE ENERGY CRISIS

The breakdown did not come all at once—not like the cataclysmic night-fall that blacked out New York and most of the Northeast in 1965—but it was no less eerie. House lights went out; furnaces sputtered and cooled: auto traffic jammed up at darkened intersections. Dog races were canceled because the electric rabbits would no longer run. Factories shifted to a four-day week, then a three-day week, laying off 1.6 million employees. Only the most essential services operated full time—hospitals, water and sewage plants—and nobody knew how long they could continue.

A scenario for the future? Perhaps. But it all happened last winter, when Britain's coal miners went on strike for almost two months. Without coal, there was not enough fuel for electric power plants. Without enough electricity, the nation faltered.

From Time, June 12, 1972, p. 49.

Symptoms of an energy crisis are emerging in the United States and elsewhere. Critical shortages of fuel oil have developed. Natural gas producers are seeking substantial rate increases, and electric companies in some regions have been unable to supply needed power, causing brownouts and blackouts in many industrial areas. As other fuels become scarcer, some coal companies have been swamped by new orders from the utilities. The present crisis merely preludes the longer term problem of maintaining adequate energy resources.

New Sources of Energy

Given that an adequate supply of energy is required to support the current quality of life, and that the largest present sources of energy (oil and natural gas) may be largely exhausted within the next 60 years, what replacements are available? Possible future sources of energy fall into four main categories—solar, tidal, geothermal, and nuclear. We shall briefly consider the feasibility of each of these alternatives.

Solar Energy. Direct use of solar energy could theoretically meet foreseeable future energy requirements for the entire world. The annual supply of solar energy to the earth is 60,000 times greater than all the energy provided each year by burning fossil fuels. Unfortunately, there are no economically feasible ways foreseen to tap this energy. Collection of solar energy for power would probably be restricted to desert areas between 35 degrees north and south latitude, which receive large amounts of sunshine. Even in these regions, present technology and schemes proposed for harnessing solar energy would require huge areas of land on which to collect the energy in amounts significant to be considered useful. For example, M. King Hubbert estimates that the area required to produce electrical power equal to United States consumption in 1970 is 24,500 square kilometers (somewhat less than one-tenth the area of Arizona). The area required to produce 1000 megawatts (equivalent to modern electrical power stations) is 42 square kilometers or a square area 6.4 kilometers per side.

Direct use of solar energy for electrical production is possible, but technical difficulties of great complexity exist, and the high cost of the required equipment makes large-scale production impractical. Some scientists, as well as science-fiction writers, have proposed the construction of orbiting satellites equipped with large areas to collect solar energy and beam it by microwave to collecting stations on the earth as a means to tap this power source. A five-mile square area of solar cells in stationary orbit above the equator could transmit enough power to supply New York City. The receiving antenna on the ground would be an open mesh of wires six miles in diameter. Economical construction of such equipment would depend upon a relatively low-cost space shuttle service for fabricating large structures in space. However, this type of energy production is not likely to be accomplished in time to meet present or impending energy shortages.

Tidal Energy.   Plans for large-scale future use of energy either from *water power* (hydroelectric) or from the harnessing of *ocean tides* appear impractical. The total potential water power capacity of the world is estimated to be about 4 times as great as presently installed electrical power capacity, but one must consider that harnessing water power impinges upon many natural scenic areas. In addition, reservoirs and dams constructed at these sites are continuously silting in from sediment and consequently have a useful life expectancy of only 100 to 200 years. Tidal power obtained from harnessing the energy of seawater as it fills and empties partially enclosed coastal basins can, in some areas, be accomplished. Dams are constructed to enclose these basins so that the difference in water level between high and low tide can be used to drive turbine electrical generators. Several tidal electrical plants are in operation in France and the USSR, but relatively few coastal areas are suitable for the harnessing of tidal power, and world potential production of energy from this source seems extremely small. On the other hand, although its total energy-yielding potential is insignificant compared to world demand, utilization of tidal power in suitable local areas poses relatively little threat of environmental pollution compared to other means of power production.

Geothermal Energy.   Can we harness heat from the earth's interior to produce power? Geothermal power can be obtained from the extraction of heat stored in large reservoirs of hot water that fill sands and deep sedimentary basins within the earth's crust. Wells drilled into such reservoirs of superheated water or steam conduct steam to the surface, where it is used as an energy source to drive conventional steam electrical power plants. Geothermal power plants presently exist in Italy, northern California, and New Zealand, and new plants are under construction in several areas.

Scientists at the Los Alamos National Laboratory are investigating a plan to derive energy from the hot (300° C+) granite underlying the earth's crust. According to the plan, two holes 15,000 feet deep would be drilled. Cold water pumped down through one hole would be heated to steam by the hot rocks and would rise in the second hole. It is estimated that steam harnessed in this fashion could supply United States energy needs for the next 3000 years.

Barring such as yet unproved technological breakthroughs, M. King Hubbert estimates that the world's potential for geothermal power represents only about 20 per cent of the present total electrical power capacity of the United States. He concludes that "while geothermal energy is capable of sustaining a large number of small power plants in a limited number of localities, it still represents only a small fraction of the world's total energy requirements, and this for only a limited period of time."

Nuclear Energy.   Can our final alternative for future power sources supply the world's growing population? Nuclear energy is derived from the atom by two primary processes — fission and fusion. Fission involves the splitting of heavy atoms such as uranium-235; fusion is the combining of light nuclei of atoms such as deuterium (see box insert Chap. 6). In general, the heat derived from these nuclear reactions is used to produce steam to drive turbines for the production of electricity. All nuclear reactors now in operation or being planned in the United States depend upon the splitting or fission of uranium-235 or certain other heavy nuclei (see Chap. 7). Electrical production from this type of reactor is growing rapidly in the United States, doubling at present every 2.4 years. In mid-1972 the United States had 25 nuclear power plants in operation, 53 more under construction, and 59 on the drawing boards.

However, the supply of uranium-235 in the earth's crust to fuel these reactors is not large. The United States Atomic Energy Commission (AEC) estimates that with reactors constructed according to current technology, domestic resources of uranium (at reasonable prices) to meet projected requirements for nuclear electrical plants are adequate only for about the next 25 years. Since the time of that estimate (1967), projections of nuclear power plant capacity have increased significantly. Thus, reserves may not last as long as the predicted 25 years. In mid-1972, nuclear power provided only about 2 per cent of United States energy needs, but it is predicted that by 1990 it will supply more than 50 per cent of these needs.

Fast Breeder Reactors.   Faced with a predicted shortage of uranium-235, several countries have pursued an intensive effort to develop fast breeder reactors. In these reactors, uranium-235 initiates a fission process that converts nonfissionable isotopes such as uranium-238 into fissionable materials such as plutonium-239. The breeder reactor creates more fissile material than it consumes, and this material is then drawn off to become a starting material for other nuclear power plants. With the breeder reactor, one gram of uranium-238 produces an amount of heat equivalent to the burning of 2.7 metric tons of coal or 1.9 metric tons of crude oil. The potential energy obtainable from breeder reactors, utilizing the much more abundant nonfissionable supplies of uranium and thorium, is hundreds to thousands of times greater than that from all the fossil fuels combined. However, conversion to breeder reactor energy production must take place before supplies of the required uranium-235 are exhausted by conventional fission reactors. If the supply of uranium-235 were depleted within the next 20 years and none remained for use in breeder reactors, it would, as M. King Hubbert suggests, "constitute one of the major disasters in human history."

In order to meet the impending energy crisis, the AEC plans to spend as much as 4 billion dollars before 1986 on a crash program to develop the liquid metal fast breeder reactor (LMFBR). Nevertheless, there is growing opposition to development of the breeder reactor among both environmentalists, who cite the dangers of the plutonium fuel used in the reactors, and economists, who question the necessity of such large-scale spending on the program. The breeder reactor has alarmed environmentalists because of its production of plutonium (the material of atomic bombs) and the use of highly reactive liquid sodium as a coolant inside the reactor. Thomas B. Cochran, a physicist conducting a study for Resources for the Future in Washington, DC, disputes the AEC's economic claims for the LMFBR as well as its contention that uranium will be in such short supply as to require a switch from conventional to breeder reactors before the end of the century. Both Cochran and the AEC agree that development of fast breeder reactors is necessary to postpone the depletion of uranium resources; but Cochran suggests that the most economically attractive choice for the remainder of this century may be the development of another type of breeder reactor, the high temperature gas reactor (HTGR), now being investigated in the United States.

Nuclear Fusion.   In 1939, Hans Bethe published his theory concerning the nuclear fusion reactions responsible for the release of tremendous amounts of energy from the sun and other stars. Such thermonuclear fusion is the basis for the functioning of the hydrogen bomb, and most recently is being considered as a possibility for almost unlimited sources of controlled energy production. Probably the most promising nuclear fusion reaction is the fusion of deuterium and tritium atoms, which releases large amounts of heat energy per unit mass (see Box Insert, Chap. 6).

Deuterium, a heavy isotope of hydrogen, is found in very minute concentrations in all water. Utilization in fusion reactions of about 1 per cent of the deuterium available from the huge quantities of ocean water could release about 500,000 times more energy than that of the world's initial supply of fossil fuels.

Development of controlled nuclear fusion energy is still in the experimental stage. Small experimental reactors have been built or are planned in France, Germany, Italy, Japan, the United States, and the USSR, but many technical problems stand in the way of producing significant amounts of fusion energy. For example, the temperature required for the deuterium-tritium fusion can reach 46 million °C. Because no material can contain such temperatures, current experiments involve confining the high temperature *plasmas* within a magnetic field. Estimates on how long it will take to develop workable nuclear fusion power vary. A few optimists propose development by 1990; some pessimists say it cannot be accomplished at all. However, many scientists believe that fusion power may be available in appreciable quantities sometime after the year 2000.

Physicists Lowell Wood and John Nuckolls of the University of California Lawrence Livermore Laboratory recently reviewed the current state of controlled thermonuclear reactor technology and suggested that less well known unconventional techniques of fusion power may be practical in the near future. They are particularly optimistic that the new technology of lasers (high energy synchronized wavelength light beams) could be used to compress small pellets of liquid deuterium-tritium to extremely high densities approaching the great densities that exist only within the centers of some unusual stars, and to fusion temperatures, thus producing an efficient thermonuclear microexplosion. The heat generated from a series of these microexplosions could be used to energize the boiler of a conventional steam-powered electrical plant and generate between 100 to 1000 megawatts of electricity. Similar but more advanced laser fusion power plants would effectively burn pure deuterium fuel (or heavy water) extracted from the ocean. Wood and Nuckolls estimate that if governmental support of research continues, a laser fusion microexplosion system to generate net electrical energy could be constructed soon.

Improvements in technology could both increase the efficiency of energy production and decrease much of the present waste. For example, a new system based on *magnetohydrodynamics* (MHD) could produce electricity directly from a high velocity flow of hot ionized gases with 60 per cent efficiency compared to the present figure of 35 per cent for electrical generation, thus providing more efficient generation of electricity with less pollution. Instead of copper wire armatures as in conventional rotary generators, MHD uses a high velocity stream of hot conductive gases in a long rocket-like tube surrounded by electromagnets. The so-called closed cycle MHD employs a high velocity stream of liquid metal or helium gas, mostly circulating in a closed coil of pipes or tubes and heated by a nuclear energy source. Supercold ($-160°$ C) superconducting power lines have been proposed that could reduce losses of energy during transmission. Wasted air conditioning, office lights that burn all night, and inefficient refrigerators that could be improved to use 50 per cent less power are some immediate (although perhaps small) sources of wasted energy that could be eliminated.

Let us return to our earlier question: can we supply enough energy to maintain an acceptable quality of life for the world's expanding population? Energy optimists Alvin M. Weinberg and R. Philip Hammond of the United States Oak Ridge National Laboratory believe there is no clear limit to the amount of energy that man can acquire and that "the limit to population set solely by limits to energy production is very large indeed—considerably larger than 20 billion [people]." If this prediction is true, the real problem arising from increasing energy production will be pollution.

Fusion power would produce less air pollution than conventional fossil (coal and oil) burning power plants and less radioactive waste than present nuclear fission power plants. It would probably also be safer from serious accident than present nuclear power plants. Nevertheless, some radioactivity would probably escape to the environment.

Any type of energy production releases heat into the environment, causing problems of thermal pollution (see Chap. 7), which could arise on a global scale. To produce enough energy for 20 billion people, technologists envision large nuclear parks located on the seashore or even offshore where the ocean could be used to dissipate the heat and eventually radiate it to the atmosphere (Fig. 9-13).

It has been estimated that by the year 2000, a power economy based on breeder reactors would produce 720,000 kg of radioactive plutonium. Such huge amounts of radioactive material could pose serious problems of nonpolluting safe storage. Some radioactive wastes must be isolated for 600 to 1000 years in order to render them biologically harmless, and long-term storage of the large quantity of waste presents significant problems (see Chap. 7). Thus, even if current energy sources are not exhausted and new sources are found, the

Figure 9-13 Artist's conception of an agro-industrial complex of the future, in which the energy of the atom is used to transform an arid desert region into productive farms and cities by supplying water, fertilizer, industrial chemicals, metals, etc. The usable portion of the earth's surface could be more than doubled in this way. (From Weinberg, A. M., and R. P. Hammond: Limits to the use of energy. Am. Sci. *58*:416, 1970.)

production of energy cannot continue indefinitely, because resulting pollution and environmental degradation will set limits to further growth.

Combining efficient, relatively low polluting energy production methods on an individual household basis may offer a solution to the energy crisis. For example, a research team at Cambridge University is currently building a model house that can generate its own power, independent of outside sources. Solar energy will be used to provide heat, rain will provide water, sewage and food and garden wastes will generate methane gas for cooking, and recycling will be used to reduce heat and water losses. Wind power from windmills will generate electricity.

WHAT WILL BE THE EFFECT OF UNITED STATES POPULATION GROWTH ON RESOURCES?

As already pointed out, the United States puts more pressure on resources and the environment because of high economic productivity and high waste than any other nation in the world. Furthermore, projections indicate that increases in population will produce tremendous increases in the demand for resources.

The Commission on Population Growth and the American Future examined the demand for nineteen major nonrenewable minerals and predicted that, with certain qualifications, the United States should have no serious difficulties in acquiring the supplies of minerals it needs for the next 50 years, even if the population were to grow at the three-child rate. Minor difficulties may include an increase in price and the mining of immense quantities of lower grade minerals. The commission's optimistic projection, however, should be qualified by several possible contingencies. First, if impending shortages are not accurately anticipated, prices may not rise soon enough to stimulate the necessary changes in economy. Second, large expansions in mining operations may result in heavy pollution and could conflict with environmental policy. Finally, existing imbalances in access to resources could lead to international problems and power struggles between regions that control valuable minerals.

During a recent fuel shortage, worldwide government attempts to decrease energy consumption included restrictions on office hours, lunch hours in restaurants, auto driving (only certain days and speeds allowed), attendance at recreational games and facilities, and even television viewing hours. If adequate amounts of water, food, minerals, and energy are not available, resource shortages will become frequent and more acute and human freedom will be increasingly restricted.

REFERENCES

Barnes, J.: Geothermal power. Sci. Am. *226*(1):70–77, 1972.

Boerma, A. H.: A world agricultural plan. Sci. Am. *233*(2):54–69, 1970. (The United Nations Food and Agriculture Organization plan to close the gap between food production and population growth by 1985.)

*Borgstrom, G.: The Hungry Planet: The Modern World at the Edge of Famine. New York, Collier Books, 1967 (paperback). (The gap between population and food, and the possibility of mass starvation within 50 years.)

Energy crisis: are we running out? Time, June 12, 1972, pp. 49–55. (A general nontechnical review of the problem and alternatives.)

Gough, W. C., and B. J. Eastland: The prospects of fusion power. Sci. Am. *224*(2):50–64, February 1971. (Recent advances bring fusion-power close to the level of scientific feasibility.)

Hammond, A. L.: The fast breeder reactor: signs of a critical reaction. Science *176*:391–393, 1972. (Reviews the problems and shortcomings of the liquid-metal fast breeder reactor development as outlined by T. B. Cochran.)

Hammond, A., W. Metz, and T. Maugh II: Energy and the future. Washington DC, American Association for the Advancement of Science, 1974.

Holt, S. J.: The food resources of the ocean. Sci. Am. *221*(3):178–194, 1969. (A well managed world fishery could significantly increase the present ocean harvest.)

Hubbert, M. K.: The energy resources of the earth. Sci. Am. *224*(3):60–70, 1971. (Discusses sources of energy, and points to the brevity of fossil fuels as an energy source.)

Lappé, F. M.: Diet for a Small Planet. New York, Ballantine Books, Inc., 1971.

*Recommended further reading

(paperback). (How to use vegetable protein currently wasted, especially in the United States—includes excellent recipes for balanced vegetable protein meals.)

Lessing, L.: New ways to more power with less pollution. Fortune, November 1970, p. 78. (New technologies of superconduction, magnetohydrodynamics, and solar power could cleanly increase power.)

Meadows, D., et al.: The Limits to Growth. New York, Universe Books, 1972 (paperback). (An important report on the Club of Rome's Project on the Predicament of Mankind, using computer systems dynamics.)

Mills, G. A.: Gas from coal: fuel of the future. Environ. Sci. Tech. 5(12):1178–1183, 1971. (With natural gas in short supply, coal gasification may provide a large source of energy.)

*National Academy of Sciences–National Research Council: Resources and Man. San Francisco, W. H. Freeman and Co., 1969. (An excellent fact-filled study with recommendations on population growth, food, mineral and energy resources.)

*Pimentel, D., et al.: Food production and the energy crisis. Science 182:443–449, 1973. (An energy shortage could threaten modern "energy intensive" agriculture and world food supplies.)

Pinchot, G. B.: Marine farming. Sci. Am. 223(6):15–21, 1970. (Several types of mariculture offer potential for food from the sea.)

Pirie, N. W.: Orthodox and unorthodox methods of meeting world food needs. Sci. Am. 216(2):27–35, 1967. (With changes in attitude many presently untapped sources of food could increase the world supply.)

Rienol, R., and L. Rienol: A Moment in the Sun, New York, Ballantine Books, Inc., 1967 (paperback). (Report on deteriorating quality of the United States environment because of pollution.)

Starr, C.: Energy and power. Sci. Am. 225(3):37–49, 1971. (Energy consumption has grown rapidly and now poses problems.)

Weinberg, A. M., and R. P. Hammond: Limits to the use of energy. Am. Sci 58:412–418, 1970. (The limit to population set by energy is extremely large.)

Wood, L., and J. Nuckolls: Fusion power. Environment 14(4):29–33, 1972. (Reviews current fusion technology and is optimistic about laser-fusion power.)

Young, G.: Dry lands and desalted water. Science 167(3917):339–343, 1970. (Desalting seawater offers the prospect for expanding areas of arable land.)

Zero Population Growth: A Teacher's Guide to Materials on Population. Denver, Colorado, ZPG, PO Box 18291, 1971. (An excellent source of further materials on population and resources. Includes bibliography of books, audio-visual materials, and population games.)

*Recommended further reading

PROBLEMS

1. What is the average depth of the ground water in your area? Are there any problems with pollutant contamination? Has the depth level fallen because of heavy usage in recent years? (Data may be available from local government agencies.)

2. Describe any instances of resource shortages which have occurred in your local area in the past year.

3. Keep a record of food and beverages you consume for several days, and from tables of caloric and protein values calculate your average daily intake. Are you receiving adequate calories and protein for your age, sex, and activity level?

4. What is meant by a "nonrenewable" resource?

5. Of the four primary energy alternatives to the fossil fuels (solar, tidal, geothermal, and nuclear), which do you be-

lieve represents the best future possibility? If it were utilized on a large scale, can you envision any problems that
might occur?

6. Keep a notebook beginning in the morning and list every
   occasion at which your activity consumes some resource.
   Looking over your notes, list ways in which the level of consumption or usage could have been decreased.

7. Average per capita energy consumption differs greatly between countries. What factors contribute to such differences?

BRAINSTORM

It is the year 2000. The Ministry of Power has announced
the following:
1. Scheduled blackouts will continue nationwide for an
   undetermined time. Brownouts will also be in effect
   during evening hours beginning April 22.
2. Licenses for the use of any new home appliances will
   not be issued until further notice from the MP.
3. All electrically amplified instruments, which were first
   banned in 1984 as nonessential, must now be turned
   over to the National Artifacts Museum. These include
   the electric knife, can opener, razor, toothbrush, dishwasher and air conditioner, which were banned from
   public use in 1995.

Adapted from ZPG, 1971.

WORLD GAME

Directions for World Game

Divide the class into seven geographic groups based on the
1970 percentage of the world population, and send each group
to a separate area of the classroom. Thus, in a class of 50
students, group 1 (Asia) will have 28 students, group 2 (Africa)
8 students, and so forth.

Now take out a bag containing the world's total "resources"
(about 50 marshmallows are a favorite choice). Distribute to
each group its "share" (percentage) of the resources.

Repeat the exercise assuming it is the year 2000. Follow up
with a class discussion of the significance of resource distribution.

10

THE PRESENT AND FUTURE CHALLENGE

Technology has badly outraced the political and social means of handling the problems it generates.

P. Siekevitz

The task is clear. The task is huge. The time is horribly short. In the past, we have had science for intellectual pleasure, and science for the control of nature. We have had science for war. But today, the whole human experiment may hang on the question of how fast we now press the development of science for survival.

John Platt

HAS TECHNOLOGICAL MAN A FUTURE?

Where are we going? What does the future hold? Is the development of science and technology leading man toward a golden age of prosperity or toward a crowded and degraded environment of ecological catastrophe?

In previous chapters we examined the rapid exponential growth of science and technology; the expansion of man's horizon that is the universe; the growing technology of biological engineering; growth in man's powers of warfare; the growing ability to control and manipulate the brain, consciousness, and behavior; the growth of man's impact on the environment, and consequent pollution; growing numbers of people in the world; growing consumption of resources; and the rapid growth in man's ability to communicate information and to process data. We have seen that the most important characteristic of the present environment is an accelerating rate of change. Man's ability to survive as a species will be determined by how rapidly and appropriately he can adapt his values and institutions to this change.

Scientists such as John Platt now view the increasing number of world crises as a result of one large crisis of *change* — "the crises of transformation." The rate of change is accelerating. We are on an exponential curve of technological growth. Within the past 100 years speed of communication has increased 10^7 times the rate of a few thousand years ago; speed of travel 10^2 times; speed of handling data 10^6 times; energy resources 10^3 times; power of weapons 10^6 times; ability to control disease 10^2 times; and rate of population growth 10^3 times. People who are now about thirty-five years old have grown up in a world totally different from that of their parents.

In recent years interest in the future has grown rapidly. Within the past decade, research into the future has been conducted by institutions such as the Ford Foundation's Resources for the Future, the Rand Corporation, the American Academy of Science's Commission on the Year 2000, and the Hudson Institute directed by Herman Kahn. The purpose of this research is not merely to predict but to chart out alternative futures on which to base plans. By constructing these alternatives, planners hope to affect the shape of the future by decisions they make today.

In their book, *The Year 2000*, Herman Kahn and Anthony Wiener identify a complex long-term "multifold trend" consisting of 13 interrelated historical trends that will continue to the year 2000 (Table 10–1). They then construct baseline statistics on such variables as population, literacy, gross national product, energy resources, military strength, and other factors that will determine possibilities for the future of any society. From these they extrapolate to construct "surprise free" projections. Then from various hypothetical scenarios they create possible

TABLE 10–1   Basic Long-Term Multifold Trends, 1970–2000*

1. Increasingly sensate cultures (e.g., empirical, this worldly, secular, humanistic, pragmatic, utilitarian, contractual, epicurean or hedonistic.
2. Bourgeois, bureaucratic, "meritocratic," democratic (and nationalistic?) elites.
3. Accumulation of scientific and technological knowledge.
4. Institutionalization of change, especially research, development, innovation, and diffusion.
5. Worldwide industrialization and modernization.
6. Increasing affluence and (recently) leisure.
7. Population growth.
8. Urbanization and (soon) the growth of megalopolises.
9. Decreasing importance of primary and (recently) secondary occupations.
10. Widespread literacy and education.
11. Increasing capability for mass destruction.
12. Increasing tempo of change.
13. Increasing dispersal of the multifold trend to underdeveloped nations.

*Adapted from Kahn and Wiener, 1967.

alternative futures. Admittedly, the fact that any day could provide a crisis situation or historical turning point decreases our ability to foretell future events. For example, from the vantage point of the prosperous years between 1900 to 1914, it would have been extremely difficult to predict the following thirty years of struggle, warfare, and hardship that took place in Europe. Nevertheless, future alternatives need to be examined.

In 1967 Kahn and Wiener listed the 100 technical innovations most likely to occur in the last third of the twentieth century (Table 10–2). Some of their predictions, such as inexpensive home video recording and playing systems, automated grocery and department stores, and general use of automation and cybernation in management and production, are already coming true (see Chap. 3); others, such as human hibernation for relatively extensive periods, and chemical methods for improving memory and learning, are yet to materialize although they may do so before the turn of the century.

Included in their predictions for the year 2000 are increased average world affluence and leisure although they claim that the relative gap between the industrial countries and the underdeveloped countries will have "widened abysmally." They see continuing urbanization and growth of megalopolises. By the year 2000, according to Kahn and Wiener, roughly one-half of the total United States population will be concentrated in three megalopolises, which they call "Boswash" (from Boston to Washington), "Chipitts" (from Chicago to Pittsburgh), and "Sanssans" (from San Francisco to San Diego). They foresee a decreasing importance of primary (fishing, forestry, hunting, agriculture, and mining) and secondary occupations (the processing of products of primary occupations) and an increasing importance of tertiary and quaternary occupations (services in a service economy). Other aspects of the multifold future trend include increasing literacy and education, increasing capability for mass destruction, increasing tempo of change, and an increasing dispersal of this multifold trend to other less developed nations.

Kahn and Wiener list some new technologies of warfare that may be created in the final decades of the twentieth century. These include simple nuclear long-range inexpensive missiles that could be used by small nations, various kinds of death rays, many techniques for effective chemical and biological warfare, doomsday machines, tsunami (tidal wave) producers, climate changers, earth scorchers, several types of psychological or even direct mental warfare, and new techniques for disguised covert nonphysical or anonymous warfare.

Considering the enormous potential for the use of computer systems, it seems certain, as Kahn and Wiener point out,

TABLE 10-2   Probable Technical Innovations 1967-2000＊

1. Multiple applications of lasers and masers (microwave amplification by stimulated emission of radiation) for sensing, measuring, communication, cutting, heating, welding, power transmission, illumination, destructive (defensive), and other purposes.
2. New airborne vehicles, including ground-effect machines, VTOL (vertical takeoff and landing) and STOL (short takeoff and landing), superhelicopters, giant and/or supersonic jets.
3. New sources of power for fixed installations (e.g., magneto-hydrodynamic, thermionic and thermoelectric, and radio-activity).
4. New sources of power for ground transportation (e.g., storage battery, fuel cell, propulsion or support by electromagnetic fields, jet engine, turbine).
5. Major reduction in hereditary and congenital defects.
6. Extensive use of cyborg techniques (mechanical aids or substitutes for human organs, senses, limbs, or other components).
7. New techniques for preserving or improving the environment.
8. Inexpensive design and procurement of "one of a kind" items through use of computerized analysis and automated production.
9. Controlled and/or supereffective relaxation and sleep.
10. Three-dimensional photography, illustrations, movies, and television.
11. Widespread use of nuclear reactors for power.
12. New and possibly pervasive techniques for surveillance, monitoring, and control of individuals and organizations.
13. Some control of weather and/or climate.
14. Practical use of direct electronic communication with and stimulation of the brain.
15. Human hibernation for relatively extensive periods (months to years).
16. New, more varied, and more reliable drugs for control of fatigue, relaxation, alertness, mood, personality, perception, fantasy, and other psychobiological states.
17. Capability to choose the sex of unborn children.
18. Genetic control and/or influence over the "basic constitution" of an individual.
19. General and substantial increase in life expectancy; postponement of aging; and limited rejuvenation.

Table 10-2 continued on the opposite page.

that the computer can be viewed as the most "basic tool of the last third of the twentieth century." Individual computers, or at least consoles or input devices, will become essential equipment for home, school, and office, and the ability to use the computer will be more widespread than the ability to play bridge or drive an automobile. By the year 2000 we may expect, in the area of cybernation and automation, such things as moderately priced robots to do standard household chores, and information processing that will make available on a global scale resources such as the Library of Congress. The future use of computer surveillance systems is one of the most frightening aspects of the new technology. It may be possible to devise a computer alphabet on automobile license plates that could be scanned by

TABLE 10–2 Probable Technical Innovations
1967–2000 (Continued)*

20. More extensive use of transplantation of human organs.
21. Permanent inhabited undersea installations and perhaps even colonies.
22. Extensive use of robots and machines "enslaved" to humans.
23. Chemical methods for improving memory and learning.
24. Simple inexpensive home video recording and playing.
25. Inexpensive high-capacity, worldwide, regional, and local (home and business) communication (perhaps using satellites, lasers, and light pipes).
26. Practical large-scale desalinization.
27. Shared-time (public and interconnected?) computers generally available to home and business on a metered basis.
28. Flexible penology without use of prisons (by use of modern methods of surveillance, monitoring, and control).
29. New biological and chemical methods to identify, trace, incapacitate, or annoy people for police and military uses.
30. New and possibly very simple methods for lethal biological and chemical warfare.

Less Likely but Important Possibilities

1. Practical use of sustained fusion to produce neutrons and/or energy.
2. Artificial growth of new limbs and organs (either in situ or for later transplantation).
3. Suspended animation (for years or centuries).
4. Direct input into human memory banks.
5. Direct augmentation of human mental capacity by mechanical or electrical interconnection of the brain with a computer.
6. Major rejuvenation and/or significant extension of vigor and life span to 100 to 150 years.
7. Chemical or biological control of character or intelligence.
8. Verification of some extrasensory phenomena.
9. Production of drug equivalent to Huxley's *soma* (in *Brave New World*).
10. Some direct control of individual thought processes.

*Summarized and selected from Kahn and Wiener, 1967.

a television camera and automatically transmitted to a central computer bank, which would search its memory for any past violations by the driver of the automobile. Conversations could be put under surveillance by computer. When certain key words such as "subvert," "revolution," or "oppose" were heard, the computer could activate a machine to record the conversation and transmit it to central storage. Thus, one consequence of the technological explosion is the challenge to human privacy not only from computer data banks and electronic surveillance but also from such products of technology as the supersonic transport plane, which has the capability of intruding on the privacy of millions through its sonic boom.

Along with automation and mass production has come

mass education or what has been termed *technication* — the dehumanization of education, as illustrated by student protesters with IBM cards pinned to their jackets.

Among technologies with the greatest future impact will undoubtedly be the laser. High energy laser beams are already used in carrying out surgery, in measuring the distance to the moon and other extraterrestrial objects, and in guiding "smart bombs" (see Chap. 6). In the near future, we may see use of lasers extended into the mining of coal and other minerals and into military use as death rays and anti-ICBM rays. They will be used in data storage devices to reduce the size of information resources and put an entire library of 20,000 volumes on an 8 by 10 inch piece of nickel foil. The probable availability of wide channel laser communication systems will result in what is sometimes called the "information rich" society and have a profound impact on the style of life by the year 2000. Using "holographic" techniques, by the year 2000 three-dimensional television will probably be available.*

New techniques in bioengineering will provide an increased life span and better health for many people. Once the body's reaction to antibodies and "foreign" tissue is understood, progress will be made in transplantation of limbs and other body organs from one person to another and from dead to live people. Artificial mechanized limbs and electronic substitutes for organs and senses, including sight and touch, may be developed as new materials become available. Drugs will be tailor-made to act specifically on a particular biochemical reaction or pathway in the body instead of the broad spectrum now used that often produces harmful side effects. Medical statistics as well as each individual's genetic constitution may be stored on computer memory banks.

New technologies of birth control may reduce population to below the predicted 7 billion by the year 2000. The development of long-term birth control injections or timed-release pills will provide extended immunity to fertilization in women. Eventually it may be possible to block fertility for a lifetime and then induce it for a desired specified period by pill or injection.

G. R. Taylor in his book *The Biological Time Bomb* predicts a major revolution in society as a result of biological and technological advances. He claims that many of these will occur within the lifetime of persons presently middle-aged and all of them within the lifetime of those now young. According to Taylor, society will be able to cope with some of the problems generated by new biological technologies. These include, for

---

*The holograph is a photographic record of an interference pattern between the reflected light waves from an object and a second wave of interfering light. When a hologram on a film is illuminated, the original object is reconstructed in three-dimensional form in a virtual and real image.

example, control of the sex of offspring, storage of eggs and sperm, and creation of artificial placentas. However, four other technologies may challenge the very structure of society: raising of intelligence; lengthening of life span with consequent changes in family relations; indefinite postponement of death and its resulting economic and population problems; and gene alteration and its possible political consequences could all lead to a major crisis in human values (Table 10–3).

Science and technology may continue to expand man's horizon and provide material benefits for years to come, but an increasing number of scientists now believe that we are approaching limits to growth that may lead to a decline in the quality of life.

ARE WE APPROACHING AN END TO GROWTH?

Many of the problems discussed in this book result from a single phenomenon—growth. Population growth impinges on the supply of resources; growing resource consumption produces environmental degradation and pollution; and all occur in a finite world that does not itself grow. Many different elements are increasing simultaneously in this finite system, and they are all interrelated in complicated ways—population growth depends on food supply, food production depends upon agricultural investment, investment (capital) requires resources, and the use of resources produces pollution, which interferes with the growth of both population and food. These quantities are related in *feedback loops* (see Chap. 3). Feedback loops can be either positive or negative. In the feedback loop system controlling population growth (Fig. 10–1), the positive feedback loop on the left accounts for population growth. The negative feedback loop controlling population growth has the opposite effect; that is, it tends to reduce the size of the population.

In a population with constant fertility (birth rate), the larger the population, the more babies will be born each year; the more babies, the larger the population the following year, and so forth. An increase in the size of a population with constant average mortality (death rate) results in more deaths per year; more deaths leads to fewer people and fewer deaths the following year. Decreasing the negative loop of deaths leads to a rapid exponential growth of population through the positive feedback loop of births. Similarly, removal of the positive loop leads to a rapid decline of the population to zero because of an excess of deaths in relation to births. This set of reactions represents a simple feedback loop, but in a natural system,

TABLE 10–3   Technological Developments Foreseen by Panelists Consulted by the Institute for the Future*

	1970	1980	1990	2000	2010	2020	2030	Later	Never
Demonstration of desalination plants capable of producing water economically for agriculture.									
Commercial availability of a large number of new materials for ultralight construction.									
Widespread use of automobile engines, fuels, or accessories enabling operation without harmful exhaust.									
Widespread existence of regional high-speed transportation systems.									
Availability of reliable weather forecasts 14 days in advance for local areas.									
Laboratory demonstration of continuously controlled thermonuclear power.									
Economical disposal of solid wastes, or laws inhibiting use of products that do not decay.									
Techniques of cultivating the ocean that yield at least 20 per cent of the world's calories.									
Routine use of ballistic suborbital transports for passengers and cargo.									
Manned interstellar exploration to a radius of at least 24 light-years from solar system.									

*Technological developments are charted, with developments in the physical sciences above and developments in the biological sciences opposite. Each polygon represents estimates made by half of the group; estimates by the quarter of the panel that foresaw an earlier date and the quarter that chose a later one are excluded. Peak of each polygon indicates the median date when the panelists forsaw a 50 per cent chance that the event would occur. (From Brooks, H., and Bowers, R.: The assessment of technology. Sci. Am. 222(2):18–19, 1970. Copyright © 1970 by Scientific American, Inc. All rights reserved.)

Table 10–3 continued on the opposite page.

hundreds of factors interact at once, and their responses all affect one another. In order to study the complex behavior of such dynamic systems, scientists developed the field of *system dynamics*. The following description demonstrates the utility of this new analytical tool.

World Dynamics

One of the most important and far-reaching studies of this century was recently undertaken by an international team of scientists and scholars headed by Dennis Meadows of the Massachusetts Institute of Technology and sponsored by the Club of Rome, an informal international association of scientists and scholars from 25 nations. The team examined five basic factors that control and ultimately limit growth on this planet—popu-

TABLE 10–3 Technological Developments Foreseen by Panelists Consulted by the Institute for the Future* (Continued)

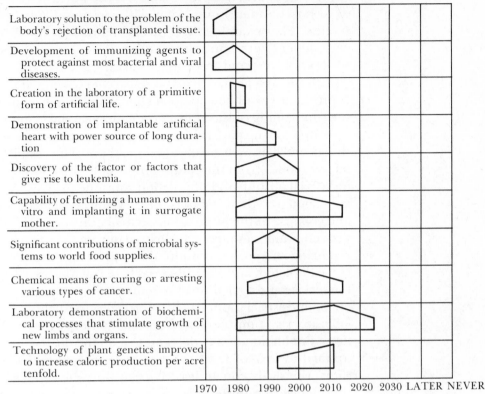

	1970	1980	1990	2000	2010	2020	2030	LATER	NEVER
Laboratory solution to the problem of the body's rejection of transplanted tissue.									
Development of immunizing agents to protect against most bacterial and viral diseases.									
Creation in the laboratory of a primitive form of artificial life.									
Demonstration of implantable artificial heart with power source of long duration									
Discovery of the factor or factors that give rise to leukemia.									
Capability of fertilizing a human ovum in vitro and implanting it in surrogate mother.									
Significant contributions of microbial systems to world food supplies.									
Chemical means for curing or arresting various types of cancer.									
Laboratory demonstration of biochemical processes that stimulate growth of new limbs and organs.									
Technology of plant genetics improved to increase caloric production per acre tenfold.									

lation, agricultural production, natural resources, industrial production, and pollution. These five basic quantities are interrelated in complex feedback loops and their interaction can best be studied using system dynamics.

The Meadows team listed the important causal feedback relationships among the five factors and quantified each, applying the best available data on global quantities. Using a computer, they then calculated the simultaneous operation of all these relationships over time from the year 1900 to 2100. The graphs produced by the computer are not exact predictions but

Figure 10–1 (Figures 10–1 to 10–4 from Meadows, D. H., et al.: The Limits to Growth. New York, Universe Books, 1972. A Potomac Associates book.)

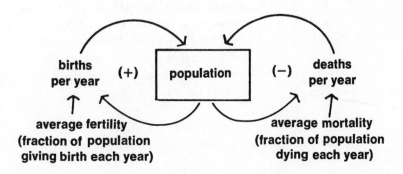

indications of the world system's behavioral tendencies for the future. They asked the question: "As the world system grows toward its ultimate limits, what will be its most likely behavior mode?" They examined several alternatives for the future, each dependent upon a different set of assumptions about how society will respond to problems arising from growth.

In the first alternative, the researchers assumed that in the future there would be no great change in human values nor in the functioning of the global population-capital system as it has operated for the past 100 years; that is, we will continue to do things much as we have always done them. The resulting "behavior mode" from the computer output indicates that food, industrial output, and population continue to grow exponentially, but growth overshoots the carrying capacity of the earth, and collapse of the population occurs because of a depletion of nonrenewable resources. Industrial output rises to a level requiring an enormous input of resources, the resources become depleted, and the industrial base collapses, taking with it the service and agricultural systems that have become dependent on industrial input (such as fertilizers, pesticides, hospital laboratories, computers, and especially energy for mechanization). Population finally decreases when the death rate is driven upward by lack of food and health services (Fig. 10–2). In short, the study concludes, "We can thus say with some confidence that under the assumption of no major change in the present system, population and industrial growth will certainly stop within the next century, at the latest."

Assuming new discoveries of resources and advanced technology that could double the resources available, an end to growth and sudden collapse would nevertheless occur. In that case, the study group found that levels of pollution resulting from overloading of the natural absorptive capacity of the environment would lead to an abrupt rise in the death rate owing to pollution and lack of food. In fact, only major changes in the present way of doing things will preclude failure of the world system into a dismal, depleted existence.

As the Meadows study indicates, technological optimism is probably one of the most common and widespread misconceptions of human society. Examining our recent history of growth, many people expect that new technological breakthroughs will continue to raise the limits to growth. However, Meadows's study indicates that even the simultaneous application of the most advanced technologies envisionable still would not prevent the collapse of civilization before the year 2100 (Fig. 10–3):

. . . Here we are utilizing a technological policy in every sector of the world model to circumvent in some way the

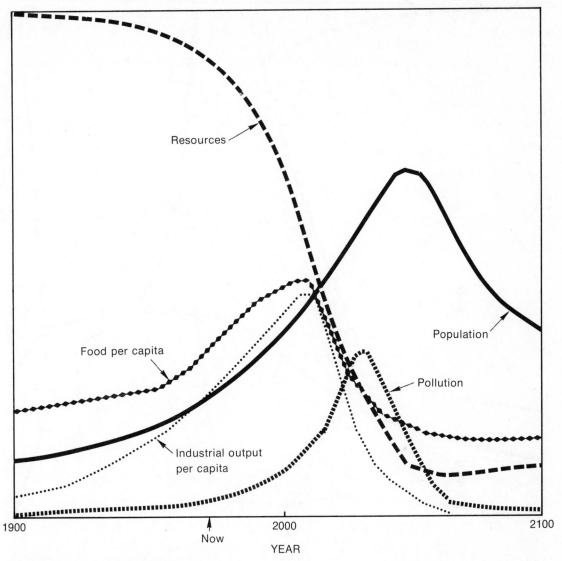

Figure 10–2 World model standard run.

The "standard" world model run assumes no major change in the physical, economic, or social relationships that have historically governed the development of the world system. All variables plotted here follow historical values from 1900 to 1970. Food, industrial output, and population grow exponentially until the rapidly diminishing resource base forces a slowdown in industrial growth. Because of natural delays in the system, both population and pollution continue to increase for some time after the peak of industrialization. Population growth is finally halted by a rise in the death rate due to decreased food and medical services.

The horizontal scale in Figures 10–2 to 10–4 shows time in years from 1900 to 2100. With the computer the progress over time of five quantities has been plotted:

population (total number of persons)

industrial output per capita (dollar equivalent/person/year)

food per capita (kilogram-grain equivalent/person/year)

pollution (multiple of 1970 level)

nonrenewable resources (fraction of 1900 reserves remaining)

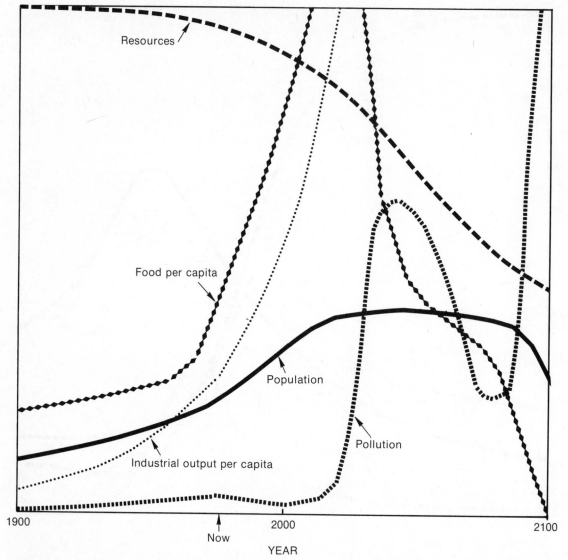

Figure 10–3 World model with "unlimited" resources, pollution controls, increased agricultural productivity, and "perfect" birth control.

Four simultaneous technological policies are introduced in the world model in an attempt to avoid the growth-and-collapse behavior of previous runs. Resources are fully exploited, and 75 per cent of those used are recycled. Pollution generation is reduced to one-fourth of its 1970 value. Land yields are doubled, and effective methods of birth control are made available to the world population. The result in temporary achievement of a constant population with a world average income per capita that reaches nearly the present US level. Finally, though, industrial growth is halted, and the death rate rises as resources are depleted, pollution accumulates, and food production declines.

various limits to growth. The model system is producing nuclear power, recycling resources, and mining the most remote reserves; withholding as many pollutants as possible; pushing yields from the land to undreamed-of heights; and producing only children who are actively wanted by their parents. The result is still an end to growth

before the year 2100. In this case growth is stopped by
three simultaneous crises. Overuse of land leads to erosion,
and food production drops. Resources are severely depleted by a prosperous world population [but not as prosperous as the present US population]. Pollution rises,
drops, and then rises again dramatically, causing a further
decrease in food production and a sudden rise in the death
rate. The application of technological solutions alone has
prolonged the period of population and industrial growth,
but it has not removed the ultimate limits to that growth.

Many technological developments—recycling, pollution
control devices, contraceptives—are vital to the future of
human society. If these are combined with deliberate checks on
growth, a world system sustainable for a considerable period of
time without collapse and capable of satisfying the basic material requirements of its people could be produced. However,
stabilizing population growth by itself is not sufficient to prevent overshoot and collapse; stabilizing capital alone is not sufficient; but constraining population and industrial growth simultaneously could produce several systems with reasonably
high industrial output per capita and long-term stability. A
stabilized world of "global equilibrium" with a sustained high
quality of life could be obtained by achieving zero population
growth by 1975 and industrial growth by 1990, and by in-
troducing the following policies in 1975: (1) resource consumption per unit of industrial output is reduced to one-fourth of its
1970 value; (2) consumption of goods shifts toward services
such as education and health facilities and away from factory-
produced material goods; (3) pollution generation per unit of
output is reduced to one-fourth its 1970 value; (4) capital is
diverted to food production although this is considered une-
conomical; (5) agricultural capital is diverted to programs of
soil enrichment and preservation; (6) the average lifetime of industrial capital is prolonged through better design, durability,
repair, and reduction of planned obsolescence. The results of
these policies would be twice as much food per person as 1970
values; a world average lifetime of nearly 70 years; and a
worldwide average industrial output per capita well above
today's level (Fig. 10–4).
     The preceding example or model is based on the most
optimistic assumption, namely, that we can suddenly and absolutely stabilize population and capital. These policies are unlikely to be instituted immediately, but if they are not instituted
in the near future, Meadows's study suggests it may be
too late to achieve an equilibrium state. If, for example, we
wait until the year 2000 to institute the stabilization policies described above, both population and industrial output per capita

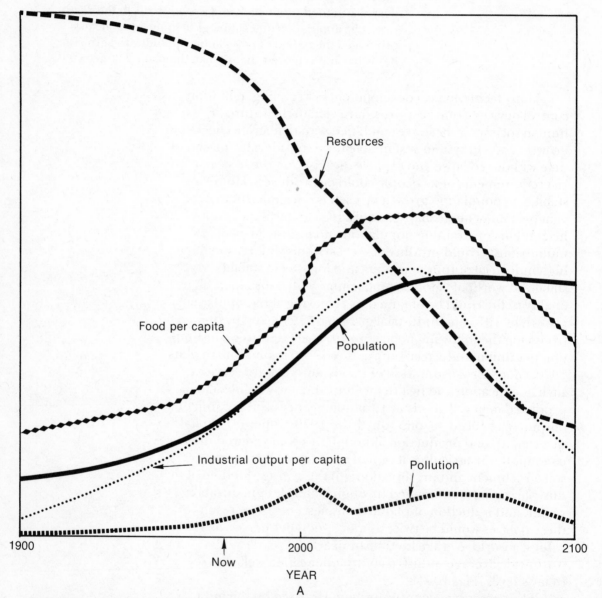

Resources

Food per capita

Population

Industrial output per capita

Pollution

1900

Now

2000

2100

YEAR

A

Figure 10–4 A, World model with stabilizing policies introduced in the year 2000. If all the policies instituted in 1975 in Figure 10–3 are delayed until the year 2000, the equilibrium state is no longer sustainable. Population and industrial capital reach levels high enough to create food and resource shortages before the year 2100.

Figure 10–4 continued on the opposite page.

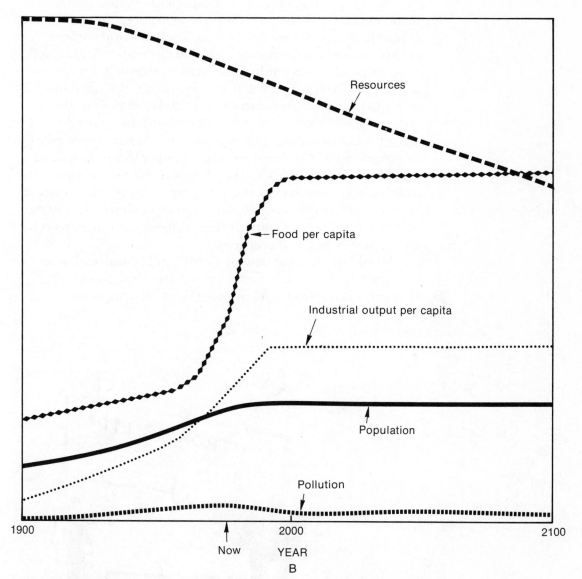

Resources

Food per capita

Industrial output per capita

Population

Pollution

1900   Now   2000   2100

YEAR

B

Figure 10-4 *Continued.*

B, Stabilized world model I.

Technological policies are added to the growth-regulating policies of the previous run (Fig. 10-4*A*) to produce an equilibrium state sustainable far into the future. Technological policies include resource recycling, pollution control devices, increased lifetime of all forms of capital, and methods to restore eroded and infertile soil. Value changes include increased emphasis on food and services rather than on industrial production. As in Figure 10-4*A*, births are set equal to deaths and industrial capital investment equal to capital depreciation. Equilibrium value of industrial output per capita is three times the 1970 world average.

would reach much higher levels, and resources would become severely depleted in spite of the conservation policies mentioned above. Food and resource shortages would occur before the year 2100, and the equilibrium state would no longer be obtainable (Fig. 10–5).

Bringing a deliberate end to exponential growth is the challenge that must be met if future generations are to inherit a habitable planet. What would life be like in a world without growth? An equilibrium state, in which population and capital are stable and forces tending to increase or decrease them are in a carefully controlled balance, does not mean stagnation. All activities would not freeze to a standstill. Meadows's study group suggests that a society no longer burdened with the many problems resulting from growth might have more energy and ingenuity available for solving other problems. Emphasis could be shifted to activities that do not require large inputs of irreplaceable resources or result in environmental degradation. For example, people could continue to enjoy desirable activities such as education, art, music, religion, basic scientific research, athletics, and social interactions.

Would the achievement of global equilibrium lead to an increase or a decrease in human freedom? Meadows's group answers this important question in the following way:

Equilibrium would require trading certain human freedoms, such as producing unlimited numbers of chil-

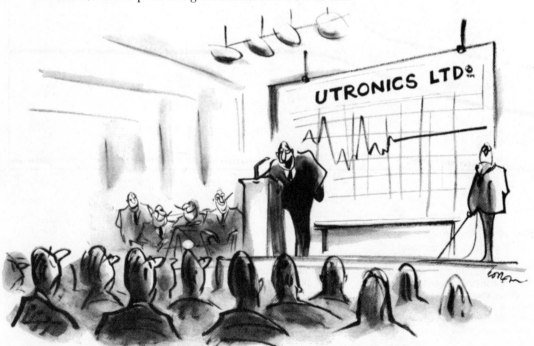

Figure 10–5 "And so we here at Utronics take great pride in announcing that we are the first American corporation to achieve zero economic growth." (Drawing by Lorenz. © 1972 The New Yorker Magazine, Inc.)

dren or consuming uncontrolled amounts of resources, for other freedoms, such as relief from pollution and crowding and the threat of collapse of the world system. It is possible that new freedoms might also arise — universal and unlimited education, leisure for creativity and inventiveness, and, most important of all, the freedom from hunger and poverty enjoyed by such a small fraction of the world's people today.

There are some serious shortcomings to the Limits to Growth study. For example, the study grouped together a number of variables as a worldwide system, when in fact growth rates of these factors in various countries and regions differ greatly. Several types of pollution are placed in a single category even though they differ in toxicity and environmental impact. In addition, computer input based on past responses cannot predict human behavior. For example, recent concern about the population explosion and pollution may have a profound influence on human behavior, causing mankind to limit growth voluntarily. One of the most serious criticisms of the study is that it does not adequately take into account man's ingenuity in finding alternate natural resources or the true rapid exponential growth of technology that might solve some of the problems it creates. As Jeremy Bray says, "The idea of physical limits is misleading, since it ignores the process of social, economic, and technological adaptation that goes on all the time, with the constant obsolescence and innovation of successive technologies."

Robert Boyd of the University of California at Davis has argued that the use of different sets of plausible assumptions can radically affect world dynamics computer model results. Boyd altered the world dynamics model to conform to a technological optimist's view, adding a new variable of "technology" and multipliers to express the effect of technology on the other variables. His equations suppose that as the quality of life declines, society will invest more in technology to improve that quality. Improved technology results in increasing productivity, which in turn raises the standard of living and eventually drives birth rates low enough so that a utopian equilibrium is reached.

Chauncy Starr and Richard Rudman have argued convincingly that technological growth is exponential. They claim that in the future, technological growth will be determined by the specific resources allocated to that growth, which in turn are allotted according to the total resources available and "the societal expectations"; that is, technological growth is proportional to societal investment in technology.

The overall growth of a technological field is composed of a series of sigmoid curves. Each curve builds on the performance level of the inventions of the previous generation. At present, the technological component of economic growth appears to be doubling every 20 to 30 years, but an overall exponential growth can be seen in many areas, for example the growth of performance or capacity of computers or in the efficiency in lumens per watt of light-emitting sources (Fig. 10–6*A* & *B*). Even with a fixed percentage of the GNP directed to new technology, exponential growth probably will occur. These considerations, according to Starr and Rudman, "indicate that the technological component of the world simulation model proposed by Meadows and Forrester is best represented by an exponential growth function." This assumption in turn should alter the predictions made by Meadows and Forrester and reduce by several decades the time required to reach a state of equilibrium (Fig. 10–7). Thus, several new studies indicate a higher level of technological optimism than that of Meadows and Forrester's study of the limits to growth.

Furthermore, Thomas Boyle of Montreal's McGill University, after studying the computer programs used by the MIT

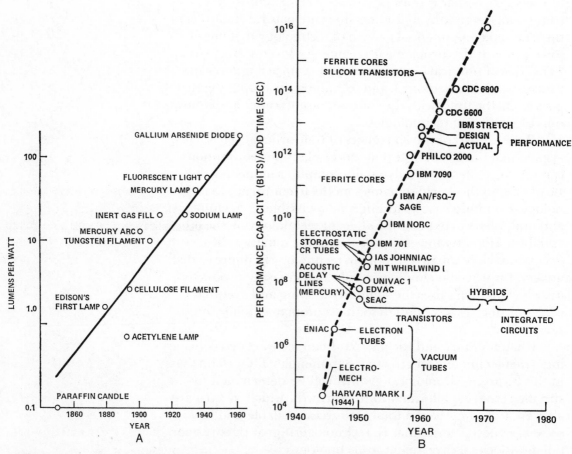

Figure 10–6   *A* and *B*. The growth of many technologies has been exponential. (Fig. 10–6 *A* from Martino, J. P.: Tools for looking ahead. IEEE Spectrum 9:34, 1972. Fig. 10–6 *B* from Bernstein, G. B., and M. J. Cetron: Technological Forecasting 1:39, 1969.)

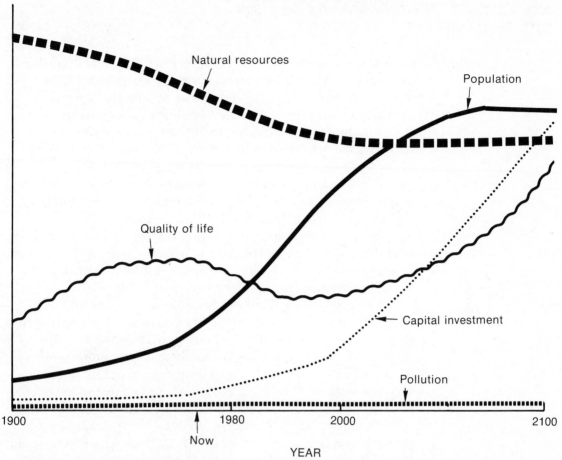

Figure 10–7 Results of the *World Dynamics* simulation altered to include the assumptions of a technological optimist. The simulation starts at 1900 and runs through 2100. (From Boyd, R.: World dynamics: a note. Science *177*:516–519, 1972. Copyright 1972 by the American Association for the Advancement of Science.)

group, claims to have found an error in the figure used for pollution levels, which he claims was assumed to be ten times greater than it should have been. Using the same assumptions as the Meadows group but correcting for the alleged error in the pollution factor, Boyle found that by the year 2100 the world population will stabilize at 6 billion people, life expectancy will rise, pollution will be under control, and technology will succeed in forestalling such crises as famine and industrial collapse. However, Meadows denies that any typographical error crept into the results, and controversy over the predictions continues.

Current world dynamics studies may be based on fallible assumptions and are only crude predictors of the future. Nevertheless, they represent a valuable new tool to be refined, mastered, and used by society as an early warning system to identify the problems generated by technology and to chart out desirable courses of action.

ARE SCIENCE AND TECHNOLOGY GETTING OUT OF HAND?

As described in previous chapters, science and technology now provide us with numerous things which we can do, but *which we must not do.* Technology is not just another historical development similar to the rise of political parties and other institutions but a new explosion in human potentialities and dangers. We have seen in recent years the emergence of what Ferry has called "the technological imperative"; that is, the belief that whatever can be done must be done. Contemporary man often accepts technological change as inevitable and irresistible, a condition described by Ellul as "technological anaesthesia." Technological innovation is accepted irrespective of harmful effects. People continue to use cars in urban areas even though such use may contribute little to mobility and may adversely affect the environment and their health.

The growth of specialized scientific knowledge and technology now challenges society's ability to understand or control its effects on everyday life. Technological decisions are now often made and carried out so rapidly that their impact on society and the environment cannot be evaluated before implementation. Harmful side effects of misapplied technology abound. Drugs are marketed publicly and later found to result in birth defects (thalidomide), cancer (tobacco), brain damage (hexachlorophene), and other ill effects. High speed supersonic transport planes are planned and constructed before their impact on the climate of the upper stratosphere or their production of noise over urban areas is adequately considered. Pesticides and other organic chemicals are spread throughout the biosphere and later are found to concentrate as poisons in the fatty tissue of humans and other animals. One of the world's largest dams (Aswan Dam) is constructed on the Nile River for irrigation, flood control, and power production, and later is found to be responsible for a widespread increase in a parasitic disease (schistosomiasis) and for the reduction in nutrients going into the Mediterranean Sea and consequent decline in fishery resources. As a result of these and other technology-related problems a growing number of people, especially among the young, have become alienated by science and technology. They have developed feelings of resentment and disaffection and feel they are subject to forces beyond their understanding and control.

Science and technology provide society with material benefits and a higher standard of living, but these are often accompanied by resource depletion, pollution, and environmental hazards. To solve these problems will require more than technological solutions: it will require fundamental changes in human values and social institutions. As Michael Baram states, "The task is to formulate coherent and humane social controls on science and technology."

At present, restrictions on science and technology operate primarily at the advanced stages of development (Table 10–4). Controls often are put into effect too late and without regard to public welfare. Legislative action (for example in suing a polluting industry) occurs only after the undesirable impact has taken place. The courts serve only as a feedback loop control and not as a more desirable feed forward control (see Chap. 3). Regulations imposed by private enterprise are usually based on profit considerations and do not consider the larger interests of conservation or public health. Citizens groups and lobbies are somewhat effective but are often understaffed and lacking in both funds and political strength. Public agencies responsible for technology control (for example, the Atomic Energy Commission) are often involved in a conflict of interest because they

TABLE 10–4   Sources of Social Control Over Science and Technology*

	Basic Science	Applied Science	Development Technology	Production, Application, and Use Technology
Scientific peer groups	X	X		
Professional associations				X
Federal government				X
executive action		X	X	X
agency programs	X	X	X	X
agency regulation	X	X	X	X
agency security				
classification	X	X	X	X
congressional hearings		X	X	X
congressional legislation				
and funding	X	X	X	X
Industry–consumer markets			X	X
Industrial associations				
and labor unions			X	X
Insurance				X
Crusaders and citizens groups			X	X
Law				
patents, copyrights,				
trade secrets			X	X
torts				X
constitutional rights				X
land use			X	X
consumer protection				X
experimentation		X	X	
Education-ethics	X	X	X	X

*Adapted from Baram, 1971.

are responsible for both promoting and controlling the same activity. Legislation to control technology carries with it no guarantee of implementation. The Refuse Act of 1899, which prohibits dumping in navigable waterways and harbors, was ignored for more than 70 years. Creation of new federal agencies and new administrative bodies to control technology again embodies the conflict of interest of experts who attempt to serve the public interest within a government agency.

Scientists as a group may have the ability to control science in its early stages of development, but they often lack objectivity, share only the limited views of their own group, and may be unable to relate their particular discipline to broad social problems. Total responsibility for science and technology cannot be entrusted to a narrow peer group of scientists. For example, it was suggested by James Watson and others that scientists in the area of genetic engineering (see Chap. 5) slow down or suspend their work and concentrate instead on the formulation of ethical controls. Scientists' responses to this idea were disturbing and include the following (Baram, 1971):

If we don't do it, somebody else will; don't worry about secret and horrible developments—all work is done in large expensive labs funded by the government; further work will improve the health of society and upgrade the gene pool; cloning of humans is at least five (or ten) years away; science is intrinsically valuable in its contribution to man's collective knowledge and it must not be controlled for social purposes of any sort.

Many scientists are approaching a new level of consciousness concerning the consequences of their work; but should the management of scientific and technological development be left to scientists alone?

It has been suggested that technology is now controlled by a few intellectual and technological "elites," who serve the private interest of small power groups. Terms such as "think tank," and "brain trust" have arisen to describe these groups.

The decisions that shaped the world of the present—at least in its physical aspects—were not the result of any painful weighing of alternatives, nor did they, for the most part, involve society as a whole. They were decisions made by individuals or groups on the basis of what, in their time and place were clear-cut and valid—but limited—goals.

Too many people suppose that restoring environmental quality boils down to a simple "search and destroy" mission against ecologic villains, or string up the ten most wanted polluters. What it does involve, is something far more difficult—it means initiating, in our society, an orderly system of making choices that has no precedent in all of human history. . . .

Charles Johnson

HOW CAN TECHNOLOGY BEST BE CONTROLLED AND DIRECTED?

Current suggestions for controlling and directing technology include proposals for (1) reorganization of federal agencies; (2) improved information flow and education for both scientists and laypersons concerning the dangers of misapplied technology; (3) improved citizen participation in the regulation and implementation of technology; (4) formation of a centralized government planning board to formulate long range goals; (5) rationing the use of national resources by citizens depending on specific environmental impact; (6) rewriting of the United States Constitution; (7) establishment of national and international advisory commissions and agencies for "technological assessment" (see later in this chapter); and (8) taxes proportional to the harmful effects of technology on population.

Social Control of Science and Technology

Baram has proposed a diversified approach for the social control of science and technology, including reorganization of federal agencies, establishment of legislative guidelines for administrative agencies (similar to EPA's environmental impact statements), fostering of independent "adversaries" citizen groups to press for legal action when necessary, and finally, changes in the present scope of school curricula. The latter suggestion probably represents the most important single social control. Present compartmentalization and specialization of graduate and even undergraduate training may need to be broadened into interdisciplinary, problem-oriented educational programs to increase awareness among scientists, pre-scientists, and nonscientists of the social impact of science and technology.

Interest is now growing in the teaching of ethics as part of a scientific education. Several private institutes have been established for the study of science, technology, and ethics. One example is Georgetown University's Institutes of Bioethics for the training of science graduate students in ethics. The unification of the disciplines of ethics and scientific research will be one way to ensure that researchers become aware during their education of the social implications of their work. In order to spread control over a broader base, Robert Jungk of the Insti-

Summary of influences and recent developments for decision-making

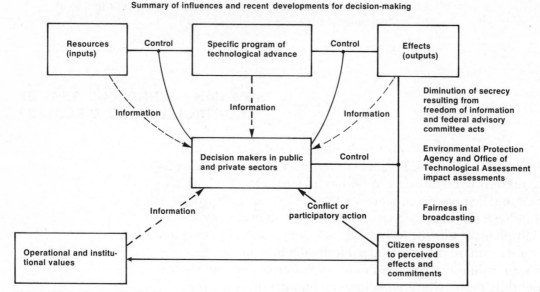

Figure 10–8 Summary of influences and recent developments for decision-making. (From Baram, M. S.: Technology assessment and social control. Science *180*:465–473, 1973.)

tut für Zukunftsfragen (Institute of Questions about the Future) in Berlin suggests the creation of "future-creating workshops," in which students and others would meet to chart out concrete proposals for alternate futures including the most desirable uses of technologies.

Public supervision of science and technology will require a clear flow of information from government, industry, and other sectors, on which citizen groups may base decisions. The 1967 Freedom of Information Act provides for public access to nonclassified government information. The Clean Air Act and Water Pollution Control Act make available to the government information often held secretly by private industries. Such mechanisms contribute to an enhanced flow of information, which can be used by citizens responding to the effects of technology. It will be an important task of corporate decision makers, the government, and the public to establish a continuing dialogue, which will support their efforts to assess and plan the uses of technology. Improved information systems such as new citizen feedback techniques from the media could contribute to this effort. Processes of information storage and retrieval in computerized systems (see Chap. 3), communications networks, and "interactive" cable television with proper safeguards against abuse could further augment citizen participation in technological decision-making.

It is often an advantage when technology is not applied too rapidly. The author recalls a conversation with one Canadian who commented on America's rush to build superfreeways to accommodate rapid automobile transportation. Many adverse environmental problems resulted, causing a further increase in the number of automobiles crowding the roads. The Canadians, on the other hand, had acted more slowly. In a certain large city, by the time they had considered all aspects of constructing a superfreeway, the harmful effects being produced by such systems elsewhere had become apparent, so they decided instead to construct a rapid mass-transit system of lower environmental impact and higher efficiency.

James Carroll of Ohio State University proposed a system of *participatory technology,* which he defines as the "inclusion of people in the social and technical processes of developing, implementing, and regulating the technology . . . when they advance a claim to a substantial and legitimate participatory role in its development . . . and implementation." Faced with the complexity of technological information, citizens as well as legislative bodies currently delegate to administrative agencies responsibility for regulating, developing, and controlling technology.

Carroll describes three major ways in which technology is currently controlled. The first method is the citizen lawsuit. Court decisions within the past few years have recognized the validity of citizen "class action" suits brought in the public interest by groups of individuals. Such lawsuits can both restrict the use of technology and lead to its modification and redevelopment into forms that better suit the public interest.

The second control on technology discussed by Carroll is *technology assessment* (see later in this chapter). Different groups view the impact of technology in different ways, as beneficial or detrimental, so that reactions and information from all segments of society are necessary for an adequate assessment. Spokesmen from many fields should have access to decision making processes. Participation and representation on a democratic basis should bring together diverse views to be considered in the actions of technological assessment. Through widespread democratic participation in technological decision making, the public can hope to obtain control of technological developments.

The third form of participatory technology described by Carroll encompasses a variety of *ad hoc* activities by individuals and groups. This form includes activist intellectualism such as writing and public speaking, quasi-official action, and political and informational activities such as lobbying groups.

Participatory technology can provide a means by which individuals who feel powerless in the face of technological complexity can find an outlet for expression of their views. However, even participatory technology could be misused. Administrative and technical officials could manipulate information in such a way as to create the illusion of citizen support for a particular course of action. Participatory technology may be important in the future control and analysis of topics described in this text including the computer and privacy, pollution control, resource exploitation, regulation of genetics and bioengineering, psychological experimentation and behavior control, and the use and handling of chemical, biological, and other technologies of warfare.

Athelstan Spilhaus, former president of the American Association for the Advancement of Science, has proposed that a *United States Planning Board* with representatives from diverse backgrounds be established to formulate long-range directions for society. Such a planning board, structured somewhat similarly to the Supreme Court, with long-term members, would generate policies in an attempt to produce "ecolibrium"—a balanced state of people with their environment. At the same time, their policy would attempt to maximize human freedom and increase individual choices. As Spilhaus points out, an increasing population and a decreasing supply of resources can only reduce the number of choices for the individual. "The great challenge is how to continue providing for people's needs and wants and yet, at the same time, to manage their environment by containing wastes . . . recycling . . . and rebuilding industry to use more efficient and less polluting techniques."

Artificially low prices breed waste and must be eliminated. Water and some other resources are priced far below their true cost. When energy is kept at artificially low prices, individual users become wasteful and industries using cheap power produce goods at less than real cost, causing consumers to use and waste them more. We can no longer think of the water and the air as inexhaustible, and just as we grow, use, and regrow food, we must now use, clean, and reclean our air and water. To the cultivation of renewable plant and fiber crops which we call agriculture, Spilhaus suggests that we must now add "atmoculture" and "hydroculture," the continuing sciences of renewing used air and water to good quality for reuse.

In a steady state world of ecolibrium, experimentation will not be absent. Indeed, new scientific and technological experimentation will be necessary to develop efficient nonpolluting ways of producing energy. Experiments in the field of sociology will be conducted to find ways to increase choices and give variety to life.

Another form of possible regulation of technology is suggested by Walter Westman and Roger Gifford, who propose an equitable system of resource allocation that could serve as an anticipatory feedback system by means of a "pricing unit." Using this system, government could control environmental impact and at the same time "provide a maximum measure of freedom for individuals and organizations." Westman and Gifford's system would begin by limiting those resource uses and activities which have the greatest and most irreversible impact on the environment. Such regulation "would require greater restrictions on personal freedom than we experience today."

In their proposed scheme, every individual would receive a rationing of *Natural Resource Units* (NRUs) expendable on a range of goods, services, and activities that have an impact on the environment. The government would establish the overall level of environmental impact for the country by fixing the total annual allocation of NRUs and the NRU price of each item, service, and activity. All individuals would be allocated an equal number of NRU's. The rich could spend their additional money on activities of low environmental impact.

To regulate population growth, children would be assigned a certain NRU value. Unlike a financial tax on childbearing, an NRU system would not financially penalize people who want to have large families. Instead, these people would simply have to weigh the desire for a large family against alternative activities, that have an impact on the environment. Industrial corporations and other organizations would also be assigned NRUs. The NRU allotment would be determined by the government on the basis of calculated allowable rates of resource consumption and pollution generation. The government NRU budget would be subjected to full public scrutiny as is its money budget.

Information and continual data would be collected for the government by an extensive computer network on existing NRU and resource reserves, population size, and pollution technology, in order to determine in what areas environmental impact must be lessened or maintained and where it may be increased. NRU currency could be held as a form of personal credit card, and spending on the account would be transferred through regional computers to central computer banks. Westman and Gifford suggest that an internationally administered body could control the environmental impact of population growth, resource utilization, and pollution on the entire biosphere. To be equitable, such worldwide NRUs would have to be allocated irrespective of national wealth. However, the authors admit that acceptance of such a scheme at the present time is highly unlikely. What would be the implications of such a system for human freedom? The authors state, "We believe the mechanism would lead to less restriction of personal freedom in a steady state society than would the current trend toward unsystematic imposition of governmental regulation."

Wilbert H. Ferry suggests that the only way to control technology in the United States is to rewrite the Constitution. He believes that political institutions and theory developed in other times offer no hope for solutions to present problems, which can only be dealt with by radically new institutions. Scientists and technologists must be "assigned to their place and the sovereignty of people reestablished. Rules must be written and regulations imposed." Because the United States is now a technological society, Ferry argues, the Constitution of the United States must be rewritten to protect citizens against the dangers of technology and to direct technology for man's welfare. However, he adds, "The constitutional direction of technology would put planning on a scale and scope that it is hard now to imagine. . . . It means working toward an integrated system. . ." Thus, Ferry's proposal, although aimed at protecting individual freedom against the technological challenge, might result instead in centralized, authoritative control, and reduction of individual freedom.

Largely a result of technology, pollution and other environmental hazards have become worldwide problems that respect no national borders. Global pollution requires a global system of research, control, and regulation by society. Whether atmospheric or aquatic, pollution produced by one country may be carried by the air, by rivers, streams, or by the sea to other countries. Only international control, regulation, and agreement will be effective in counteracting this destruction. Some progress along these lines is being made, and international pollution control agreements are being discussed by such bodies as the European Economic Community and the United Nations. Several institutions including the United States Senate, the United Nations, and an International Parliamentary Congress on the Environment have recommended the establishment of a world environment institute to deal with worldwide contamination of the biosphere.

In December, 1971, the United States Congress, recognizing the need for social control of research in biomedical technology and bioengineering, passed a resolution calling for the establishment of the National Advisory Commission on Health, Science, and Society. The commission, directed by a board of fifteen professionals from the areas of law, theology, medicine, government, and the humanities would study and evaluate the ethical, social, and legal implications of advances in biomedical research and technology.

Technology Assessment

In order to evaluate in advance the impact of new technologies, scientists are developing a new field of *technology assessment.* The United States National Academy of Sciences proposed legislation to establish a technology assessment board, which would identify social impacts of technologies and control or encourage their development. The Technology Assessment Act of 1972 created the Office of Technology Assessment (OTA) in order to "provide early indications of the probable beneficial impacts of the applications of technology and to develop other coordinate information which may assist the Congress. . . ."

Technological developments arise from a multiplicity of decisions made in industry, government, the marketplace, and by individuals. Important policies and decisions at the governmental level are becoming increasingly scientific and technical in nature. It is now the responsibility of government representatives to decide whether or not expensive programs for space exploration, supersonic transports, nuclear fusion energy, and offshore oil drilling should be instituted and to weigh possible benefits against detrimental environmental impacts. Unfortunately, these legislators often do not have the appropriate background to understand the consequences of the various technologies' applications.

The social consequences of new technologies are difficult to evaluate or predict. For example, the increase of television sets in the United States from the 1948 figure of 100,000 to the present number of about 50 million must have a large social impact. The rapid growth in automobile transportation (another product of technology), has resulted in huge lawcourt backlogs involving litigation of auto accident claims.

The National Academy of Sciences concluded that the best way to improve the assessment of technology was through the federal government. The government, they recommended, should be empowered to study and recommend policies with regard to technology but not to act; "to evaluate but not to sponsor or prevent." They recommended establishing a constellation of organizations in the legislative and executive branches of government and in the National Science Foundation.

In March, 1972, a group of scientists announced the formation of an International Society for Technological Assessment (ISTA), headquartered in The Hague, Netherlands. The society will emphasize the use of technology assessment to define problems, predict future developments, and suggest options for alternative technologies.

Michael Baram has proposed a conceptual framework model for technological assessment and social control. He points out that new technology will have its effects within the areas of health, economy, environment, resources, values, and social and political institutions and processes. The EPA requires the filing of Environment Impact Statements before major actions by industries and government are instituted. This concept could be extended to include other than environmental effects, so that all sectors of science and technology will be required to file impact statements providing information similar to that prescribed by the EPA. Such information would include (1) potential impact, (2) unavoidable adverse impacts, (3) irreversible commitment of resources, (4) short-term considerations versus long-term resource needs, and (5) alternatives to the proposed action.

Probably the best means to direct technology toward humane ends will consist of a combination of many of the methods discussed above, including improved education and information flow; more efficient government organization, assessment, and planning; and some system of equitable rationing or regulation. One thing is certain—all of these actions will result to some degree in a loss of human freedom of action.

WHAT IS THE PROBABLE LIFETIME OF A TECHNOLOGICAL CIVILIZATION?

Ecologist Barry Commoner has asked, "Are we really in control of the vast new powers that science has given us, or is there a danger that science is getting out of hand?" We have no historical precedents from which to extrapolate the probable lifetime of a technological civilization.

New technologies described in previous chapters confront society with many important problems. Basic scientific research is an expression of the human traits of curiosity, exploration, and creation (Chap. 1). It should not be abandoned, but urgent problems now cry out for an expansion of applied research that will offer specific solutions. Man will have to learn to apply wisely and humanely the knowledge he acquires. Furthermore, these urgent problems demand more than technical solutions; they require fundamental changes in human values, and they must be overcome rapidly if man is to survive for more than a few decades.

The logistic curve of growth (see Chap. 1) is now approaching its limits. Growth must cease if man is to survive and learn to live harmoniously in this new technological world. But what is the probability for human survival into the nine-hundredth generation? Leo Szilard estimated the "half-life" (the time until the chance for human survival drops to 50 per cent) of the human race to be 10 to 20 years. His prediction takes into account the probability of multiple crises. If we assume the probability of a nuclear confrontation resulting from a major political incident to be about 5 per cent (during the Cuban missile crisis, President Kennedy estimated about a 25 per cent chance of a nuclear exchange taking place); then the probability of nuclear weapons being used on a large scale before 1990 is about fifty per cent. Some agricultural experts have predicted that large-scale famines will begin within this decade, particularly in India and China, events which will lead to political instability and war.

The only chance for human survival, according to Platt and others, is an immediate "mobilization of scientists and inventive minds," government agencies, and business and labor leaders to attack crisis problems. During World War II, the technologies of sonar, radar, and atomic energy were rapidly developed by more than 10,000 scientists working in teams in the United States. We must now work with the same urgency and the same concept of a task force that were present in war-related research to solve within the next decade the problems that have been generated by new technologies coupled with rapid growth. Many scientists may have to assume new roles and turn toward solving goal-oriented applied research rather than basic research.

> Many of the creatures of the earth have seniority over us. They made it this far by remaining compatible with their environment, by adapting and adjusting to the natural circumstances of their existence. There are many species that have vanished because they could not adapt. It's not at all inconceivable that man will follow these creatures into extinction. If he continues to reproduce at the present soaring rate, continues to tamper with the biosphere, continues to toy around with apocalyptic weapons, he will probably share the fate of the dinosaur. If he learns to adapt to the finitude of the planet, to the changed character of his existence, he may survive. If not, nothing like him is likely to evolve ever again. The world will be inherited by a creature more adaptable and tenacious than he.
>
> Is there such a creature?
>
> Yes. The cockroach.
>
> From interview with Paul Ehrlich,
> Playboy Magazine, August, 1970, p. 55.

THE CHALLENGE TO FREEDOM

In 1931 Aldous Huxley predicted (in his book *Brave New World*) a world 600 years in the future in which human behavior was controlled by an advanced technology, resulting in a severe loss of freedom. In the preface to the 1946 edition of the same book, Huxley reexamined his predictions in the light of scientific and technological discoveries and advances that had taken place in the intervening years. He concluded that *Brave New World* was becoming a reality even faster than he had originally foreseen and that "the horror might be upon us within a single century."

In the 1931 edition, Huxley portrayed the Savage as a free individual, free to choose between two alternatives; an insane life in utopia, or the life of a primitive in an Indian village. By 1946, Huxley had reached the conclusion that man no longer possesses free choice. Rapid technological changes, he predicted, would result in centralized totalitarian governments arising out of economic and social confusion. Unless we decentralize and use science for humane ends, he warned, individual choice would be lost, and man would be left with only two alternatives; a nationalized totalitarianism, or a supranational totalitarianism arising from the need for efficiency and stability to counter the social chaos resulting from overly rapid technological development.

We live in a unique age — an age that departs greatly from all past history and human experience. It contains a dual future potential of both great promise and great peril. The global problems arising out of technological advancements require global solutions in the form of international institutions of law and management. However, this necessary centralized authority over resources, pollution, and growth could challenge human freedom. In this era of crises it is the responsibility of men of knowledge and scholarship to communicate the dangers and explain the choices in understandable terms to the general public and to those responsible for political action.

Human freedom is now challenged by the following dilemma: uncontrolled laissez-faire policies of technological growth could lead to man's destruction and loss of freedom; on the other hand, centralized large-scale planning and regulation to control technology and to avoid technological disaster may also lead to a loss of individual freedom. How much time do we have to meet and overcome this challenge to freedom? The time is impressively short. The hour is upon us. The challenge must be met. As Robert L. Heilbroner estimates,

> The coming generation will be the last generation to seize control over technology before technology has irreversibly seized control over it. A generation is not much time, but it is some time. . . .

REFERENCES

Abelson, P. H.: Limits to growth. Science *127*:1, 1972. (Short criticism pointing out some shortcomings of Meadows's study.)

Baram, M. S.: Social control of science and technology. Science *172*:535–539, 1971. (Education and other means should be encouraged to formulate humane social controls on science and technology.)

Baram, M. S.: Technology assessment and social control. Science *180*:465–473, 1973. (It is time to develop a coherent framework for the social control of science and technology.)

Branscomb, L. M.: Taming technology. Science *171*:972–977, 1971. ("Mankind must

react rationally to the opportunities as well as the problems created by technology."
Promotes wise use of technology.)

Bray, J.: Growing strong. Environment *14*(4):43–45, 1972. (A review and criticism of
Limits to Growth.)

*Brooks, H., and R. Bowers: The assessment of technology. Sci. Am. *222*(2):13–21,
1970. (National Academy of Sciences recommends federal mechanisms for
controlling or advancing alternative technologies.)

Callahan, D.: Profile: Institute of Society, Ethics and the Life Sciences. Bioscience
21(13):735–738, 1971.

Carroll, J. D.: Participatory technology. Science *171*:647–653, 1971. (A structure for
citizen participation in controlling technology could reduce alienation.)

Crow, B. L.: The tragedy of the commons revisited. Science *166*:1103–1107, 1969.
(Major problems have neither technical nor political solutions. Extensions in
morality are not likely.)

Delaying doomsday. Time, October 15, 1973, p. 47. (Report on Thomas Boyle's
purported discovery of an error in the MIT *Limits to Growth* computer program.)

Ellul, J.: The Technological Society. New York, Alfred A. Knopf, Inc., 1964.

Ferry, W. H.: Must we rewrite the Constitution to control technology? Saturday
Review, March 2, 1968. (The United States Constitution should be rewritten to
control adverse impacts of technology on society.)

Forrester, Jay W.: World Dynamics. Cambridge, Massachusetts, Wright-Allen Press,
1971. (System dynamics applied to a computer model of worldwide growth.)

Kahn, H., and A. J. Wiener: The Year 2000: A Framework for Speculation on the Next
Thirty-three Years. New York, Macmillan Publishing Co., Inc., 1967. (Examines
economic, scientific, and technical changes likely to occur by the year 2000 and
possible variations from standard scenarios.)

*Meadows, D. H., et al.: The Limits to Growth. New York, Universe Books, 1972
(paperback). (Report for the Club of Rome Project on computer prediction of
worldwide ecosystem dynamics and man.)

Platt, J.: What we must do. Science *166*:1115–1121, 1969. (Accelerating change and
multiple crises require a mobilization of active minds if man is to survive.)

Siekevitz, P.: Scientific responsibility. Nature *227*(5265):1301–1303, 1970. (Scientists
should assume responsibility of explaining consequences of their findings to the
public.)

Starr, C., and R. Rudman: Parameters of technological growth. Science *182*:358–364,
1973. (Technological growth is exponential in most fields — this fact will alter
conclusions of Meadows's study of the *Limits to Growth*.)

Taylor, G. R.: The Biological Time Bomb. New York, World Publishing Co.
(New American Library), 1968. (New biological technologies may result in a crisis
in human values.)

*Toffler, Alvin: Future Shock. New York, Random House–Bantam Books, 1960
(paperback). (Technology has created an accelerating pace of change, which now
challenges man's ability to cope.)

Westman, W. E., and R. M. Gifford: Environmental impact: controlling the overall
level. Science *181*:819–825, 1973. (A rationing system based on allocation of
National Resource Units would control environmental impact while maximizing
personal choice.)

Wilson, R.: Tax the integrated pollution exposure. Science *178*:182–183, 1972. (A tax
based on the product of concentration of pollutant and public health impact on
population is suggested.)

---

*Recommended further reading

PROBLEMS

1. Of the technological developments listed in Tables 10–2
and 10–3, which two (assuming they are developed) do you
believe will have largest impact in determining what society
will be like in the next century? Why? Which two seem least
important?

2. As indicated in this chapter, scientists differ in their estima-
tions of whether or not new technologies will be created at a

fast enough rate to prevent the growth-and-collapse world model of Meadows's study. What are some possible reasons for "technological optimism"?

3.   Computer studies to predict future growth and "quality of life" have considered the world as a single system even though growth rates differ in time and between regions. What other criticisms of these studies could be added? Are such studies nevertheless worthwhile?

4.   What is meant by the term "technological anaesthesia"? Give several examples.

5.   Some examples are given in this chapter of the misapplication of technology. Describe several additional cases in which technology was applied without prior evaluation and resulted in harmful effects.

6.   What is "participatory technology"? How could it be instituted?

7.   List several advantages and disadvantages of the "National Resource Unit" scheme of rationing environmental impact.

8.   What changes in human values do you believe might improve mankind's chances for survival in this age of technology?

BRAINSTORM

Your group has been selected to serve on a national "future-creating workshop." Considering the possible future technologies described in this chapter and throughout the text, what immediate steps can your group recommend to government and individuals to avoid harmful effects of some specific future technologies?

Consider that it may be desirable to retain maximum individual freedom.

APPENDIX

Systems of measurement are used for the acquisition of knowledge of quantities in terms of standard units. The metric system, now used internationally in science, was originally established by international treaty at the Metric Convention in Paris in 1875 and has since been extended and improved. In this book, we are concerned with four of the fundamental metric units. These are:

QUANTITY	UNIT	ABBREVIATION
length	meter	m
mass	kilogram	kg
time	second	sec
temperature	degree	°C

Larger or smaller units in the metric system are expressed by the following prefixes:

MULTIPLE OR FRACTION	PREFIX	SYMBOL
1000	kilo	k
1/100	centi	c
1/1000	milli	m
1/1,000,000	micro	μ
1×10^6	mega	M
1×10^{-9}	nano	n

Length

The **meter** was once defined in terms of the length of a standard bar; it is now defined in terms of wavelengths of light. A meter is about 1.1 yards.

1 **centimeter,** cm, is 1/100 of a meter, or about 0.4 inch.

The unit commonly used to express sizes of dust particles is the **micrometer,** μm, also called the **micron.** There are one million micrometers in a meter, or about 25,000 micrometers per inch.

One **light-year** equals the distance traveled by electromagnetic radiation traveling at speed c (3×10^{10} m/sec) in one year, that is, 9.464×10^{12} km.

Mass

The **kilogram**, kg, is the mass of a piece of platinum-iridium metal called the Prototype Kilogram Number 1, kept at the International Bureau of Weights and Measures, in France. It is equal to about 2.2 pounds.

One gram, g, is 1/1000 kg.

The references in this book to a **ton** refer to the metric ton, which is equal to 1000 kg or 2205 pounds.

Volume

Volume is not a fundamental quantity; it is derived from length.

One **cubic centimeter**, cm³, is the volume of a cube whose edge is 1 cm.

One **liter** = 1000 cm³.

Temperature

If two bodies, A and B, are in contact, and if there is a spontaneous transfer of heat from A to B, then A is said to be *hotter* or at a higher temperature than B. Thus, the greater the tendency for heat to flow away from a body, the higher its temperature is.

The Celsius (formerly called Centigrade) temperature scale is defined by several fixed points. The most commonly used of these are the freezing point of water, 0°C, and the boiling point of water, 100°C.

The Fahrenheit scale, commonly used in medicine and engineering in England and the United States, designates the freezing point of water as 32°F and the boiling point of water as 212°F.

Energy

Energy is the capacity to do work. There is energy in a mule, in a moving train, in a compressed spring, in a stick of dynamite or a pound of coal or an ounce of uranium.

The metric unit of energy is the **joule**, which is defined in terms of physical work. The energy of 1 joule can lift a weight of 1 pound to a height of about 9 inches. An **erg** is a much smaller unit; there are 10 million ergs in a joule.

The unit commonly used to express heat energy, or the energies involved in chemical changes, is the **calorie**, cal. One calorie is about 4.2 joules. The energy of one calorie is sufficient to warm 1 gram of water 1°C.

The **kilocalorie**, kcal, is 1000 calories. This unit is also designated Calorie (capital C), especially when it is used to express food energies for nutrition.

The **electron volt**, ev, equals 1.60×10^{-19} joules, and the **kilowatt hour**, kwh, equals 3.60×10^6 joules.

Area

The metric unit for larger areas is the **hectare**, ha, which is equal to 1000 square meters or 2.47 acres.

A.2   CHEMICAL SYMBOLS, FORMULAS, AND EQUATIONS

Atoms or elements are denoted by symbols of one or two letters, like H, U, W, Ba, and Zn.

Compounds or molecules are represented by formulas that consist of symbols and subscripts, sometimes with parentheses. The subscript denotes the number of atoms of the element represented by the symbol to which it is attached. Thus H_2SO_4 is a formula that represents a molecule of sulfuric acid, or the substance sulfuric acid. The molecule consists of 2 atoms of hydrogen, 1 atom of sulfur, and 4 atoms of oxygen. The substance consists of matter that is an aggregate of such molecules. The formula for oxygen gas is O_2; this tells us that the molecules consist of 2 atoms each.

Chemical transformations are represented by chemical equations, which tell us the molecules or substances that react and the ones that are produced, and the molecular ratios of these reactions. The equation for the burning of methane in oxygen to produce carbon dioxide and water is:

$$CH_4 + 2O_2 \rightarrow CO_2 + 2H_2O$$

Each coefficient applies to the entire formula that follows it. Thus $2H_2O$ means $2(H_2O)$. This gives the following molecular ratios: reacting materials, 2 molecules of oxygen to 1 of methane; products, 2 molecules of water to 1 of carbon dioxide. The above equation is balanced because the same number and kinds of atoms, one of carbon and four each of hydrogen and oxygen, appear on each side of the arrow.

The atoms in a molecule are held together by chemical bonds. Chemical bonds can be characterized by their length, the angles they make with other bonds, and their strength (that is, how much energy would be needed to break them apart).

The formula for water may be written as H $\overset{\textstyle O}{\diagup\diagdown}$ H, showing that the molecule contains two H-O bonds. The length of each bond is about 1/10,000 of a micrometer, and the angle between them is 105°. It would require about 12 kcal to break all of the bonds in a gram of water. These bonds are strong, as chemical bonds go.

In general, substances whose molecules have strong chemical bonds are stable, because it is energetically unprofitable to break strong bonds apart and rearrange the atoms to form other, weaker bonds. Therefore, stable substances may be regarded as chemically self-satisfied; they have little energy to offer, and are said to be energy-poor. Thus, water, with its strong H-O bonds, is not a fuel or a food. The bonds between carbon and oxygen in carbon dioxide, CO_2, are also strong (about 1.5 times as strong as the H-O bonds of water), and CO_2 is therefore also an energy-poor substance.

In contrast, the C-H bonds in methane, CH_4, are weaker than the H-O bonds of water. It is energetically profitable to break these bonds and produce the more stable ones in H_2O and CO_2. Methane is therefore an energy-rich substance and can be burned to heat houses and drive engines.

A.3 LARGE AND SMALL NUMBERS

Scientists often use extremely large or small numbers. For example, light travels at about 30,000,000,000 centimeters per second; and the diameter of an atom is approximately 0.00000001 centimeter. It is convenient to express such numbers as **powers of ten** in the following way:

$$10^0 = 1$$
$$10^1 = 10$$
$$10^2 = 10 \times 10 = 100 \qquad 10^{-1} = 1/10 = 0.1$$
$$10^3 = 10 \times 10 \times 10 = 1000 \qquad 10^{-2} = 1/100 = 0.01$$
$$10^{-3} = 1/1000 = 0.001$$

.

.

.

.

$$10^{15} = 1{,}000{,}000{,}000{,}000{,}000 \qquad 10^{-15} = 0.000000000000001$$

Thus 3,200,000,000 is written 3.2×10^9, and 67/100,000 (0.0000067) becomes 6.7×10^{-6}.

Numbers such as 10^6 are called **exponential numbers**, where 10 is the base and 6 is the exponent. Numbers in the form 6.02×10^{23} are said to be written in **scientific notation**.

INDEX

Numbers in *italics* indicate illustrations; numbers followed by a (t) refer to tables.